Research Reports Esprit

Subseries PDT (Product Data Technology)

Project 5168 · CACID

Edited in cooperation with
the Commission of the European Communities and
Product Data Technology Advisory Group (PDTAG)

Esprit, the Information Technology R&D Programme, was set up in 1984 as a co-operative research programme involving European IT companies, IT "user" organisations, large and small, and academic institutions. Managed by DGIII/F of the European Commission, its aim is to contribute to the development of a competitive industrial base in an area of crucial importance for the entire European economy. The current phase of the IT programme comprises eight domains. Four are concerned with basic or underpinning technologies, and the other four are focused clusters aimed at integrating technologies into systems. The domains are software technologies, technologies for components and subsystems, multimedia systems, and long-term research; the focused clusters cover the open microprocessor systems initiative, high-performance computing and networking, technologies for business processes, and integration in manufacturing.

The *Esprit Research Reports* series is helping to disseminate the many results – products and services, tools and methods, and international standards – arising from the hundreds of projects, involving thousands of researchers, that have already been launched.

Springer
Berlin
Heidelberg
New York
Barcelona
Budapest
Hong Kong
London
Milan
Santa Clara
Singapore
Paris
Tokyo

R. F. Schmidt   M. Schmidt (Eds.)

Computer Aided
Concurrent Integral
Design

Springer

Volume Editors

Rolf F. Schmidt
Fachhochschule Offenburg
Badstr. 24, D-77652 Offenburg, Germany

Martin Schmidt
Daimler Benz AG, FT-Austauschgruppe (TEA)
D-70546 Stuttgart, Germany

Cataloging-in-publication data applied for

Die Deutsche Bibliothek - CIP-Einheitsaufnahme

Computer aided concurrent integral design / R. F. Schmidt and
M. Schmidt (ed.). - Berlin ; Heidelberg ; New York ; Barcelona
; Budapest ; Hong Kong ; London ; Milan ; Paris ; Santa Clara
; Singapore ; Tokyo : Springer, 1996
  (Research reports ESPRIT : Project 5168, CACID ; Vol. 1)
  ISBN 3-540-60480-4
NE: Schmidt, Rolf F. [Hrsg.]; Research reports ESPRIT / 5168

CR Subject Classification (1991): J.6, K.6.1, H.4.0, H.2.4

ISBN 3-540-60480-4 Springer-Verlag Berlin Heidelberg New York

Publication No. EUR 16886 EN of the European Commission,
Dissemination of Scientific and Technical Knowledge Unit,
Directorate-General Telecommunications, Information Market
and Exploitation of Research,
Luxembourg.

© ECSC-EC-EAEC, Brussels-Luxembourg, 1996
Printed in Germany

Neither the European Commission nor any person acting on behalf
of the Commission is responsible for the use which might be made
of the following information.

Typesetting: Camera-ready by the editors
SPIN: 10486583   45/3142-543210 – Printed on acid-free paper

Foreword

Product Data Technology encompasses the information related to all stages in the product life cycle from product design via production planning, production processes, production control etc. through the delivery and operational stages of the technical product. Product Data Technology takes a coherent, unified view of the information captured in this whole life cycle and provides methodologies to support this integrated perspective.

The Product Data Technology Advisory Group (PDTAG), a Special Interest Group of the CIME Division of the Commission of the European Communities founded in 1991, has encouraged the formation of a new subseries on Product Data Technology within the existing series of Research Reports ESPRIT. This subseries will provide a depository for the important contributions made by ESPRIT projects to the evolving area of Product Data Technology, particularly also those based on STEP (ISO 10303). PDTAG is grateful to Springer Publishers for establishing this subseries which will serve to report recent international developments in Product Data Technology.

The current volume describes the results of the ESPRIT Project CACID (Proj. No. 5168), dealing with the subject of computer-aided concurrent integral design. Concurrent engineering provides an important enhancement of computer-aided product design by offering a framework and methodology for the management of design projects with multiple parallel tasks. Concurrent Engineering is one of the important growth areas for Product Data Technology. It presents new paradigms for design process management for the synchronization of several parallel tasks with the aim of shortening development cycles while lowering cost and securing quality. The CACID Project has made valuable contributions to this rapidly developing subject. It has developed a new conceptual platform and realized a practical prototype by integrating the capabilities of product modeling, acquisition and usage of standard parts and the support of concurrent design. The results of CACID thus illustrate the current status and future promise of concurrent design systems.

Horst Nowacki
Chairman of PDTAG

Partners of the CACID ESPRIT Project:

Albert Nestler Electronics GmbH & Co., Lahr, Germany (Nestler)
(Coordinator)

Institut für Rechneranwendung in Planung und Konstruktion,
Universität Karlsruhe, Germany (RPK)

B.E.R. DESSINDUS S.A., Colmar, France (Dessindus)

Association Française de Normalisation, Paris, France (afnor)

Falko Standard GmbH, Wien, Austria (FAST)

Tecnation per Innovazione Technologica S.P.A., Torino, Italy (Tecnation)

Authors

Dr. Rolf F. Schmidt, Hans Marggraff (Nestler)
Karl Hain, Martin Schmidt (RPK)
Hubert Freyburger, Jean-Claude Heitzler (Dessindus)
Jacques Le Quéré (afnor)
Nicolaus Ondracek (FAST)

Preface

In autumn 1989 the CAD development department of Nestler electronics in Lahr and the research institute for computer usage in planning and design in Karlsruhe reflected on co-operating in the development of a future CAD system. The researchers from the RPK, very well acquainted with the theories of mechanical engineering, very much liked the user-friendly designer-minded NesCAD system. Nestler was interested in building more designer's knowledge into the next generation of their CAD systems. It was considered to launch together an ESPRIT project. The intention was to enlarge the usage of conventional CAD systems from the detail design phase to embodiment design, which so far was hardly supported by information technology. Thus both needed a third partner with the practical knowledge of daily mechanical engineering design.

Nestler was able to interest one of its customers, the French mechanical engineering company Dessindus in Colmar, in the project. Dessindus had the urgent need to reduce their development times in embodiment design and had already performed some experiments in concurrent work on drawing boards. Consequently the idea of 'Computer Aided Concurrent Integral Design' (CACID) was born. Dessindus insisted on the importance of standard solutions selectable by technical and design criteria and we were able to convince the French standardisation association, afnor, which is also a NesCAD customer, and the Austrian software house FAST, which has a lot of experience in standard parts subsystems, to participate in the project. There were negotiations with another Spanish industrial partner, whose information technology department was very interested in the project, but they could not win the manufacturing department over to evaluate the project results. At last Tecnation from Torino, Italy joined the project to perform market studies to ensure a wider application of project results in the European market.

The project was favoured by the ESPRIT evaluators and started in October 1990. It became very successful, yielding new insights in special fields of product data technology as well as delivering practical results realised in the design process. The usage of CAD systems was expanded to preliminary design phases. Concurrent design in the embodiment design phase became reality. Real exploitation started while the project was still running. The open system architecture designed and implemented in CACID as well as the communication and user interface tool kits form the base of a novel product family, developed in close co-operation of Nestler, strässle and rwt, which supports concurrent engineering in the fields of CAD, CAM and CAQ.

I would like to thank the people who played key roles in the project for their enthusiastic and co-operative work on the project. First of all Dr. S. Bridge from the RPK (who meanwhile has moved to the European Patent Office) and R. W. Mayer from Nestler, who together worked out the technical annex, which settles most of the novel CACID ideas. Furthermore B. Malle, M. Schmidt and K. Hain from RPK, H. Marggraff from Nestler, H. Freyburger and J.-C. Heitzler from Dessindus, N. Ondracek from FAST, M. Juliard and J. Le Quéré from afnor and P.G. Motta from Tecnation. Now we expect and hope that the products derived from the project will succeed on the market.

Lahr, January 1996   Rolf F. Schmidt

Table of Contents

1 Introduction

1.1 Industrial Need for Concurrent Design

Life times of today's products get shorter and shorter. This means that new products with a better functionality or new design replace older products on the market in shorter periods of time. In contrast to this decreasing period of time on the market, the development period increases because of modern products' complexity and the number of involved experts. To solve this problem the concept of Concurrent Engineering has been introduced. According to Penell and Winner [14] Concurrent or Simultaneous Engineering can be defined as

> A systematic approach to the integrated, concurrent design of products and their related processes, including manufacture and support. This approach is intended to cause the developers, from the outset, to consider all elements of the product life cycle from conception through disposal, including quality, cost schedule, and user requirements.

The design phase is one phase of the engineering process. In this phase designers search for technical solutions according to customers requirements. Especially the design phase of a product very often needs too much time. This is the reason why methodologies and tools which support concurrent or co-operative design are of much interest.

Designer are used to using Computer Aided Design (CAD) systems to support their work. In recent years CAD systems have evolved from simple computer graphic systems to sophisticated geometric modellers. Furthermore, current research is moving towards so called 'product modelling' systems for the management of all product related data which arises during the product life cycle. However, as yet there are no systems for the effective support of concurrent product design, i.e. design work carried out by a design team of several engineers working simultaneously on the same design project.

1.2 Objectives of the CACID Project

The objective of the CACID project is the development and implementation of a pre-production prototype CAD system for the support of concurrent design activities. An integral part of the proposed CACID system is the integration of the use of standard parts into the concurrent design process and the possibility of adding company specific standards. The CACID system should also incorporate some of the principles of methodical design by providing a mechanism for describing the product functionality in the concurrent design process.

The CACID system should support concurrent design processes by providing state of the art geometric modelling abilities which are enhanced as follows:

Additional engineering information is included to form the CACID modeller. Mechanisms for the control of concurrent design processes ensure the consistency of the CACID model even if several designers work concurrently. A conferencing tool allows the interactive exchange of information between members of the concurrent design team.

The development of the system will be subject to continual practical validation through the close involvement of industrial partners: their design engineers will support the design of the system and use it (early prototype) for real concurrent design work. Feedback from the use of the early prototype will be incorporated into the final pre-production version.

The impact of the CACID system is the definition of a new generation CAD system which is no longer dominated by the mathematics of geometric modelling but instead matches the practical needs of industrial design engineers.

1.3 Project Consortium

The project consortium includes six partners from four countries. The CAD software house Nestler Electronics GmbH (Germany), the consultant Tecnation S.P.A. (Italy), the standard part supplier Association Française de Normalisation (France), the standard part system developer Falko Standard GmbH (Austria), the special machines manufacturing company B.E.R. DESSINDUS S.A. (France) with its own design department and the CAD research institute RPK (Germany). The project is organised from the company Nestler.

1.4 Structure of the Book

The book is structured into four parts. Chapter 2 gives a short summary about the term Simultaneous Engineering and recent CAD technology. In Chapter 3 the design process found at a small design department (company Dessindus) is analysed. Based on this analysis requirements to a new type of concurrent CAD system are defined. Chapters 4 and 5 deal with the conception and implementation of the CACID system. Because the project had a duration of only two years not all ideas will be implemented. The realised prototype is evaluated by the company Dessindus. The results of the evaluation phase are documented in Chapter 6.

2   State of the Art

2.1 Concurrent Design and Aspired Benefits

Simultaneous / Concurrent Engineering and Design are new keywords for the research area and also for industrial practice. Somehow many engineering processes are already divided into several simultaneous as well as concurrent activities with the goal to shorten the period of time needed for the product development. A further goal is to increase the quality of the process itself. For a first common understanding, the terms simultaneous and concurrent engineering are defined as:

> *Simultaneous Engineering* is the engineering process which results from starting engineering process stages even if earlier process stages are not finished (figure 1).

Simultaneous Engineering stresses 'when' a process stage is started and the need for an effective information interchange in a multidirectional way. This is different from the sequential view of product development. Here normally process stages are first started when earlier stages are completely finished. This results in a 'one-way' information channel.

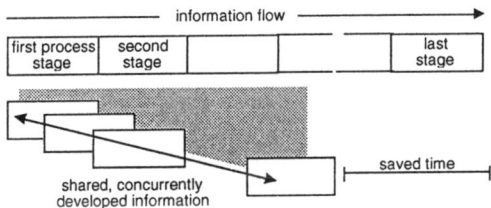

Figure 1: Simultaneous Engineering Process

The term Concurrent Engineering is often used in the same way like Simultaneous Engineering. Because the term concurrent stresses a special aspect of Simultaneous Engineering we define:

> *Concurrent Engineering* is the engineering process which results from solving an engineering task with the help of a team (figure 2).

In contrast to the term simultaneous, the term concurrent stresses 'where' is 'what' done. This means task decomposition and solution composition are important. Sometimes also the term co-operative engineering or design is used to stress that the engineers have to work closely together even if they have different tasks. The product which is designed by the team should fit in an optimum manner all user requirements, manufacturing and cost constraints.

Corresponding to the definition of simultaneous and concurrent engineering we can define Simultaneous Design as the design process which results from starting design process stages even if earlier design stages are not finished. Concurrent Design is the design process which results from solving a design task with the help of a team.

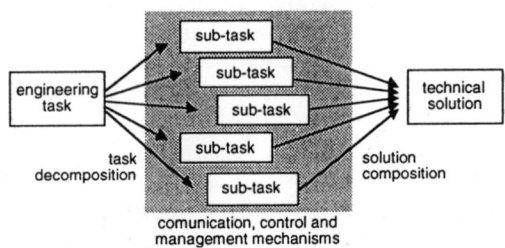

Figure 2: Concurrent Engineering Process

Because the CACID project deals with concurrent design an extended definition is given:

> Concurrent design is a systematic approach to the integrated, concurrent and simultaneous design of a product by a team. The members of the design team - which may have very different skills - must be explicitly co-ordinated and further be able to exchange information for this purpose.

Currently there is only a vague knowledge about a Simultaneous Engineering methodology. Very different disciplines play an important role in this area. These are i.e. theories about organisation, design, communication and also social aspects. But in general it seems to be clear that a methodology is needed, companies organisation should be changed and flexible tools should support information interchange and as much as possible transparency.

One Concurrent Engineering methodology was i.e. developed by M. Cutkofsky and J. Tenenbaum [3]. They developed a methodology for concurrent product and process design and a computational framework that supports it. The methodology is based on the premise that manufacturability is best assured by simultaneously designing a part or assembly and the process used to fabricate it.

An other interesting source for the CACID project was the General Architecture, Engineering and Construction Reference Model (GARM) [4] which was developed for solving problems in the area of architecture and is part of the Standard for the Exchange of Product Model Data (STEP) . With the developed model it is possible to describe complex products, their requirements, their functionality as well as their solution. To support the division of a design task the GARM describes a product as a hierarchical tree of functions and sub-functions. For every sub-function one or several technical solutions can be defined.

Aspired Benefits of Concurrent Design

According Pennell and Winner [14] the benefits of concurrent design can be grouped into three main categories:

- *Improved quality*:
  Quality improvements are reached because cross-discipline teams of engineers can detect mistakes in the product design earlier (design for manufacturing). The quality of the product and the quality of the design process itself is improved.
- *Lower costs*:
  Cost reductions is achieved by reduced costs in the design phase (the number of design interactions are reduced) and reduced costs during manufacturing (most problems are already solved in the design phase).
- *Shorter development cycles:*
  Decreased development cycles are mainly achieved by cross-functional teams. If they work together they can improve their techniques and react more flexible on problems.

Somehow it is hard to quantify the real benefits of Concurrent Design because its realisation means changes in the organisation, better information interchange and good team work. The result should not only be shorter development periods, but also an improved quality of the development process.

2.2 CAD Systems

2.2.1   Market Survey of Concurrent CAD Needs

A considerable demand for concurrent integral engineering systems has developed in recent years. The development of concurrent design systems is now generally seen to have a large market. This is underpinned by several market studies worked out by major consultancy firms. These have determined, that the concurrent engineering market is growing over proportionally during the next ten years. The demand for concurrent and integral design is driven by the pressure companies in Europe experience from oversee markets, in particular the markets of the Pacific Rim.

Today's manufacturers face continuously higher demands on their products, while at the same time they have to deliver them in always shorter life cycles. The fitness of the product for new purposes demands completely new features designed into the products. This is expressed in the 'design for' metaphor. Product development departments are required to develop new skills in order to develop products which are:

- designed for manufacture,
- designed for reliability,
- designed for recyclability,
- designed for the environment,
- designed at optimised cost,
- design for maximum customer satisfaction, and
- custom design for specific niche markets.

This demands higher quality products to be developed. In order to achieve this companies have adopted the Japanese approaches to Total Quality Management (TQM). These approaches stress the optimisation of the whole development process. The idea behind this approach is, that only a well structured quality oriented development process will yield high quality products. Companies which have long term strategies will put quality highest, since it will deliver the greatest long term benefits. Companies that implemented these ideas early are obviously earning the benefits now. Products which fulfil all these new requirements tend to be more complex than today's products. This is detrimental to achieving high quality products at optimised costs. In order to find optimal i.e. efficient solutions more engineers and designers have to work together closely. Integrated product development is seen as the solution of meeting those demands. This is ideally suited to be supported by information technology.

Companies that have taken a practical 'paper and pen' approach to integrated product development soon find themselves overwhelmed by the complexity of the data to be managed and organised. This slows down the product development process, and makes it vulnerable to errors.

Hence manufacturing companies turn to computer aided information technology in order to manage this complexity and the data that is associated with it. The technology needed for this purpose is product data modelling.

Virtually every company that is pressured to develop higher quality products in order to remain competitive has a need for computer aided information technology to implement 'Integrated Product Development'.

At the same time the product life cycle is generally reduced, and more specialised products have to be developed for niche markets. As a result more products have to be developed in a shorter time. This is considered to be a danger for the above quality goals. Reduced time to market cannot be achieved by developing primitive cheap products. This is particularly true for European manufacturers, which have always been delivering high quality.

This leads to the approach of concurrent design. The task of developing a new product is tackled with many people at the same time. These come from different parts of the company and add different skills to the design process. They work concurrently on the same product, delivering results as fast as possible. Working this way has several consequences:

Large development teams are needed to get maximum concurrency. Many tasks have to be solved at the same time. The tasks which are solved are not completely independent. They share information and modify design documents concurrently. If the access to the design documents is not organised in any way interference between sub tasks is likely to occur.

The use of product data bases can also help to solve the problems in this area. They offer mechanisms to organise and control access to design data. In order to achieve concurrent engineering customers demand tools for organised access to design documents, which allow a large number of people to access and update the design without conflict. Information technology is able to deliver this for all documents, which are stored electronically, and for which a semantic model exists.

The second point still causes the continued dissatisfaction of customers with currently available systems. The mechanisms which are offered by today's base technology do not reflect the way of working of most companies. E.g. today's relational database technology is not able to model the structure of typical design data, and access mechanisms are based on formal elementary data units and not on the units of thinking of the designers. Hence all solutions based on this technology have problems with complex structured documents, with a large number of interdependencies. Therefore research must not only concentrate on tools, but on high level semantic models of the design data. The more a model is able to capture the semantics of a design document, and the more it reflects the way of working of the involved persons, the more it can deliver concurrency and efficiency.

The expectations from the market define a system which is highly adopted to the company's working style. The computer support in product development up to now has been centred around solutions, that are only feasible with a computer like stress analysis or numerical controlled machinery etc. In these domains computer solutions were not required to compete against proven practical approaches in terms of productivity and cost.

Now in a concurrent engineering framework, all types of documents must be handled electronically, even those which are handled efficiently today without computer support. Customers will not accept a deterioration in the overall efficiency. For these types of documents techniques are required to manage them even more efficiently.

The documents produced during product development are all highly interdependent. I.e. there exist many links from one document to information contained in another. These links must be handled by a concurrent design system.

In order to realise where the benefits of concurrent integral design lie, it's important to look what happens today (figure 3). In the marketing department, requirements for a new product are worked out. Customers are requested what they would like, and of course they want it all. The document with the requirements is handed to the design department. There the document is analysed. Many points important for the design are left unspecified, and many requirements are in conflict if they are to be solved at once. The designer try to invent a product, that meets all requirements, and fill out the missing information as good as they can. Since they are unsure, they prefer the more thorough solution, just in case marketing complains.

The first delay occurs. Marketing is not consulted to solve the problems, because they will change everything again, which would require a major redesign. The design gets more complex and more expensive. The design documents are not made available to the production, because they continue to change, and taking requests into account from production people would take even longer. Only when the completed design documents are available they will be handed to the production people. These plan the production and find out that they need large investments into the production facilities in order to produce such a product. Basically they think that once again the designers have not taken care of manufacturablity.

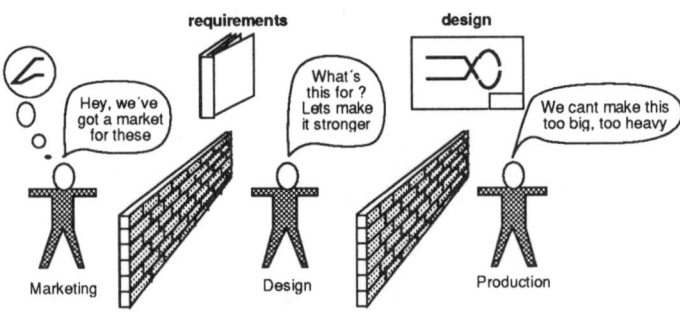

Figure 3: Factory of Today

Now that production plans are complete costs are calculated and it turns out that the product will be too expensive. Now the design goes back to the other departments. And forth again and back again. This is the situation that must be remedied.

This can be achieved by using a single document, which can be accessed by everybody involved in the product development process. Such a document is the product data model. It is used to achieve the following notions:

- Control access to the documents. I.e. who is allowed to read and update documents.
- Automatically distribute information to the people concerned.
- Store and automatically check the engineering constraints.
- Work on a design with several designers at the same time.
- Provide a unique access path for product development information, such as:
  - Which designer is assigned to which task.
  - Which alternative solutions exist for a task.
  - Which decisions have governed the selection of a particular solution.

However it is important to keep in mind, that users will not accept cumbersome overhead in the use of such a system. It is not sufficient to deliver any solution, that is capable of doing these things, but a solution must be found that delivers high productivity to the users. Users demand solutions which not only provide the necessary functionality, but which deliver the maximum productivity through ease of use, extendibility, adaptability, scalability and reliability.

The following components influence the market acceptance of concurrent design systems (figure 4):

- user constraints,
- functionality,
- enabling technology.

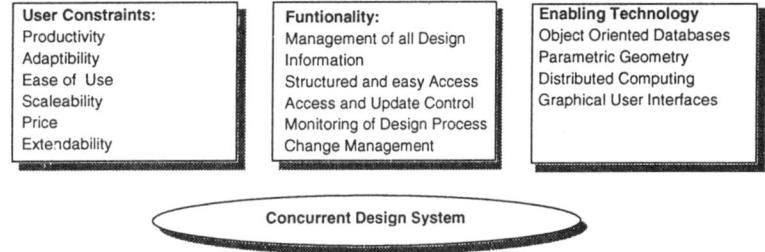

Figure 4: Factors of Concurrent Design Systems

When designing such a new system it must be kept in mind, that it meets an already functioning IT infrastructure in the companies. These companies expect that a new system fits in with the existing systems. For instance 2D CAD systems still outnumber 3D systems by a factor of 100:1. Hence users demand that the system must also be usable with 2D geometry.

2.2.2   Concurrent CAD Systems

When the CACID project started in 1990 there were only few 'concurrent' design systems on the market. Most of these systems were results from academic research groups. The system *Designers Aid for Simultaneous Engineering* (DAISIE) [1] is e.g. a knowledge-based CAD system that helps mechanical designers conduct simultaneous engineering on their conceptual and layout design. An object oriented language serves as medium for describing designs and for sharing knowledge among several experts. An other knowledge-based concurrent engineering system is proposed in [9] (Concurrent Engineering With Suggestion-Making CAD Systems). It is able to make design suggestions based on manufacturing and design knowledge. The implementation is based on a feature modeler. S. C-Y. Lu and S. Subramanyam [12] (A Computer-Based Environment For Simultaneous Product and Process Design) tured design environment based upon the knowledge-based systems concept to simultaneously perform part and process design (design for manufacturing).

A commercial available system was *Cadds4X -Assembly Design* [16] which was implemented by Prime Computer. It is able to support the simultaneous work of several designers on several detail design drawings and the general assembly drawing. Additionally the system supports the modelling of solids and sculptured surfaces.

Meanwhile a broader set of tools - so called groupware - are on the market. The needed base technology like object oriented database systems, well defined product model data exchange protocols like STEP and high-level commu-nication facilities like the Common Object Request Broker Architecture (CORBA) from the Object Management Group are now in a state that it can be used by application oriented systems. One of these systems is e.g. the SIFRAME system from Siemens/Nixdorf Informationssysteme AG.

3 Construction of a Special Machine

3.1 Introduction to the Concurrent Design Process

The company Dessindus is a typical small size enterprise with an engineering department of about ten designers and a production of about fifteen people. To improve there competitiveness they try to solve their design task quicker than other companies. This goal is reached by the extensive use of standard parts and concurrent design techniques. Their design process currently splits into ten phases:

- clarification of the design task,
- research of alternative solutions,
- development of the conceptual design,
- preliminary design,
- changes of preliminary design,
- general design,
- detail design,
- control of the detail drawings,
- correction of the detail drawings, and
- storage of the design drawings.

Currently 2D CAD systems are used within the detail design phase. As future CAD systems should also cover the general design phase, we will concentrate on these two phases in the following.

3.2 A Design Task

The given design task is to design a machine which is able to produce cleaning rags. The cleaning rags consist of two different layers of material - fibre an non-woven foil. The main design task is to design a mechanism which is able to stamp both layers together and give them a certain pattern which influence the quality of the new cleaning rag. Additionally some requirements are to be regarded in respect to the specification sheet handed over by the clients:

- certain standard parts should be used ...
- cost should not be higher then ...
- the design task should be solved until the end of ...

3.3 Conceptual and Preliminary Design

After the clarification of the design task - requirements, constraints and wishes from the customer - the first steps in the design process are part of a very creative and informal phase. To solve a design task first of all the designers develop alternative solution principles (figure 5) for the main product functions.

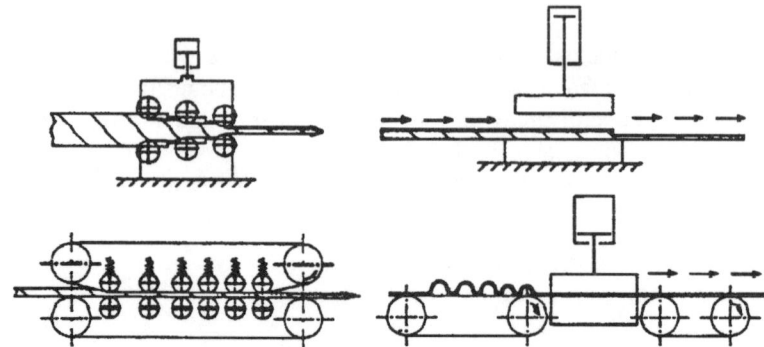

Figure 5: Alternative Solution Principles

Every proposed principle solution has several disadvantages and advantages. One (or more) of the alternative solution principles are chosen to be worked out more detailed (figure 6) to validate that the chosen principle really.

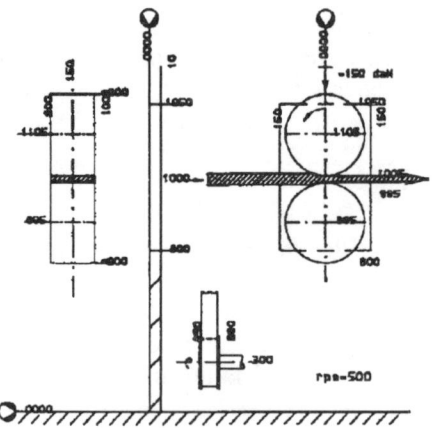

Figure 6: Proposed Solution with Main Dimensions

To detail the proposed solution, the principle solution is integrated into the machine (figure 7):

- location of the positioning stress onto the machine,
- design of the possible dimensional stress been made,
- design of all components in relation with the expected location of the part been installed, and
- design of parts been connected onto the machine.

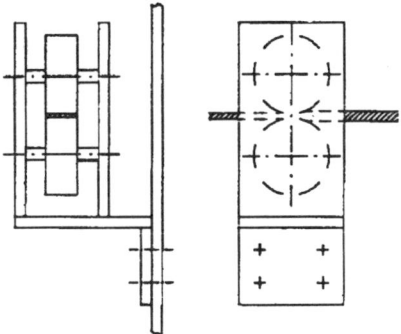

Figure 7: Preliminary Design Sketch

3.4 The General Design

When the general design phase begins, the physical principle and the solution principle is already fixed. The designer has a rough idea of what the product looks like. This is the man who usually was involved in the prior design phases. He will monitor also the whole design process. The size of the design team directly corresponds to the complexity of the task and the physical size of the object to be designed. As a rule of thumb an 'A0-size' object is done by two designers. Smaller ones are often handled by a single designer.

Based on the adopted solution, the designer or draftsman proceeds with the definition of the design space. In our calander study case one must also take into consideration the transport of the material at the input side of the calander, its evacuation at the output side, as well as the movement of the mechanical parts.

As for the transport of the product at the input and output sides of the calander we have adopted that of a double superimposed conveyors. The product passes between the two belts of the conveyors, its advance being achieved by the pinching effect. The choice of this form of the transport has been made taking into account that which exists already on the machine to be modified.

3.4.1 Definition of Design Spaces

The definition of design spaces has the goals to delimit the zones and to affect them to the different sub-assemblies to be constructed. This delimiting is easy for the designer using a sketch done in 2D from the front view and from the top view. In this approach he takes into consideration that the construction elements are interactive, the zones are thus delimited by the contours materialised by a polygon. In our calander study (figure 8) we have defined two design spaces baptised A and B:

- the design space A represents the environment attributed to the calander itself, and
- the design space B represents the environment attributed to the input an output conveyors of the calander.

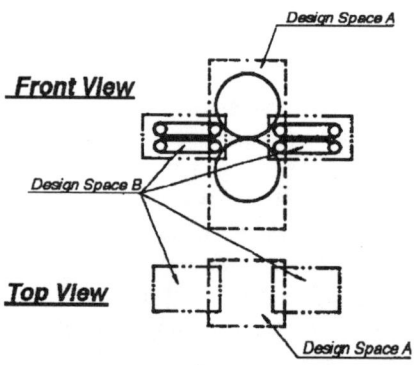

Figure 8: Definition of the Design Spaces

3.4.2 Remark Concerning Design Spaces

Our two design spaces are the delimitations of the volumes that we shall attribute to the different designers associated with this study. These design spaces define also the sub-assemblies which will be making up our calander assembly. It is to be noted that these volumes may change in size during the study. They are also interwoven, one into the other, forming common zones. At this point the project leader attributes the conception tasks to several designers (figure 9) who will work simultaneously:

- in our study case the calander sub-assembly corresponding with the design space A is attributed to the designers Dupin and Fritsch.
- the transport conveyor sub-assembly corresponding with the design space B. The transport conveyor sub-assembly could have been given to two

designers, but this presents no real interest as their conception would be similar.

The conception, one could say, starts at this point. The designer Dupin has priority as he must first define roughly the dimensions of the calander, only afterwards can the designers Degan and Fritsch start working as they are tributaries of the elements of the designer Dupin.

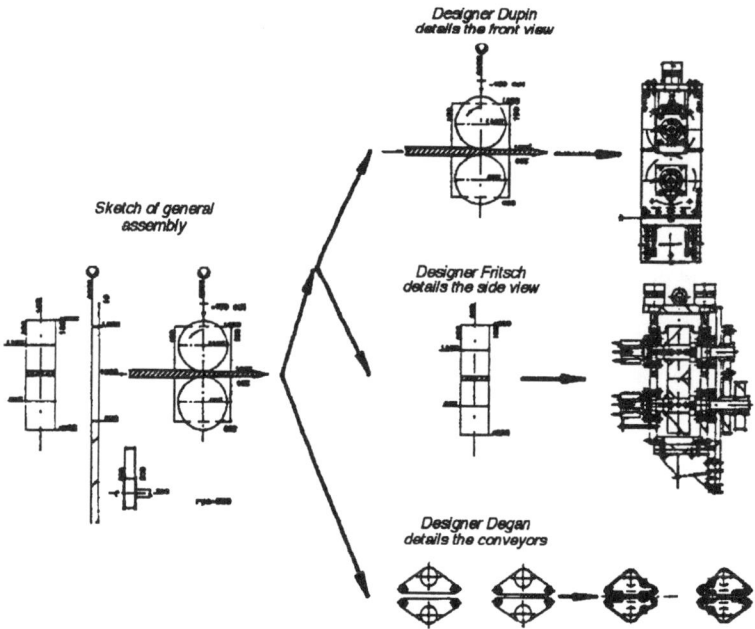

Figure 9: Concurrent Design in Different Views

The final size or the maintainability are important parameters which influence this step. The size of the product to be manufactured is also an important parameter. The second steps deals with dependent functional parts, which may be standard parts also. The size of these dependent parts is influenced by the main components, so calculations regarding the dimensions take place now. A criteria for selecting a standard part is beside others the availability within the project time.

During this step a special kind of concurrent design technique arises at Dessindus. A second designer starts with the second view of the model. After projecting the drawing of designer one into his view he starts designing the objects in his view. That means, two designers are designing an object simultaneously in two views. A strong communication between this designers is recommended. This technique allows a very early error detection because on designer controls the other designer.

Designing dependant functional parts continues the work. The integration of standard parts which may be developed outside the company, takes place now (figure 10).

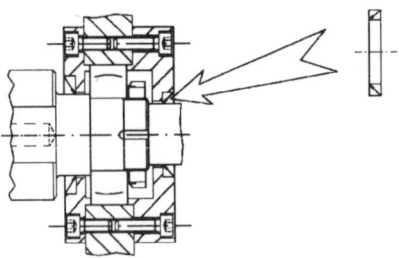

Figure 10: Integration of Standard Parts

The last design step finishes the environment of the machine. Secondary parts are added to the general assembly drawing (figure 11).

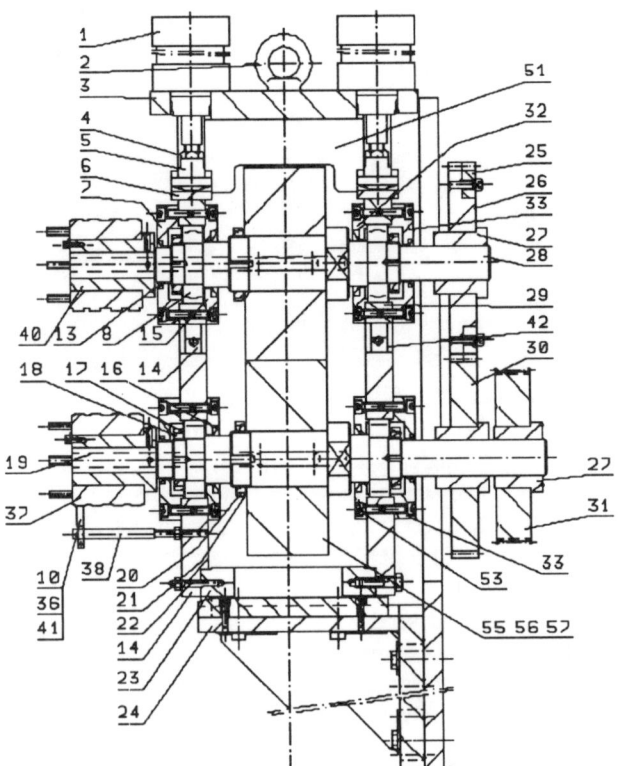

Figure 11 : General Assembly Drawing

DESSINDUS			NOMENCLATURE		Date:Decembre 90	Page 1/4	
			Sous-ensemble:CALANDRE		Plan d'ensemble N°		
N° Plan	Rep	Nb.	Désignation	Référence	Fournisseur	Form.	Observation
	1	8	Vérin double effet ø12 course 85	CUFR/XLR/X5	RACAL		
	2	1	Anneau de levage	650 tps	HAPRLEX		
6830-10	3	1	Plaque supérieure			A1	X040-Brut
	4	8	Ecrou de tige	77 170 (M16x1,5)	GP840		
6830-11	5	8	Embout de vérin			A5	X030-Brut
6830-12	6	8	Attache vérin			A3	X045-Brut
6830-13	7	1	Plaque avant extérieur haut			A2	X130-Brut
	8	4	Roulement à rotule sur rouleaux	88 808 00	SKF		(jeu C3)
	10	8	Galet fonds		BDX 6873		Fourniture
6830-14	11	1	Olavette			A3	X045-Brut
	12		Graisseur				
	13	4	Joint d'étanchéité	D.700 330/03	RMLXTRA		Type:IK (Vitaa)
6830-15	14	1	Plaque latérale avant			A1	X045-Brut
6830-16	15	1	Plaque avant intérieur haut			A2	X030-Brut

DESSINDUS				DF
				CACID

Figure 12: Parts List

Finally the parts list is prepared (figure 12).

3.5 The Detail Design Phase

During the detail design phase the completion of the components takes place. Although all dimensions are almost fixed there are still changes to the objects necessary or information to be added.

Chamfers might be added or changed to fulfil the 'inner order' of the component. These changes influence again other parts. So there is still a sort of 'micro design', which needs also communication between the designers (figure 13).

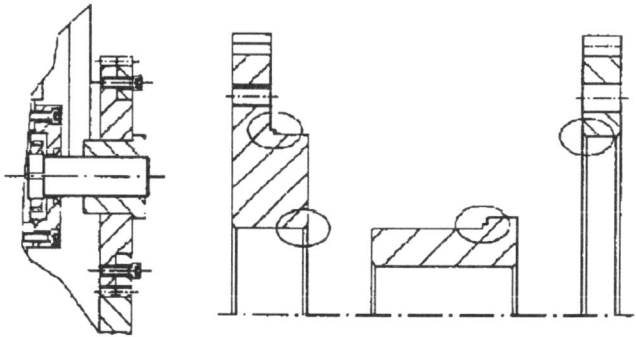

Figure 13: Problematic of Concurrent Detail Design

3.7 Requirements to a Concurrent Design Tool

At Dessindus today concurrent design is done:

* in drawing several views simultaneously,
* in designing different parts with different designers, and
* by doing general design and detail design of different assemblies of one product in parallel.

Concurrent design can be supported by several tools. The most interesting one seems to be CAD systems. Most of the actual industrial CAD systems only support drafting mechanisms.

To support concurrent design effectively the following features should be available:

* the results of the preliminary design stage must be available on the CAD system for the next design stages. There is a need for describing product functionality, solution principles, and rough shapes.
* there must be a possibility to describe design sub-tasks which are to be worked out by several designers. There is a need to support the management and organisation of design projects.
* to allow the simultaneous modification of product describing information there is a need for control and communication mechanisms among the designers. Especially:
    - geometrical consistency control between different designers, and
    - functional consistency control.

* design decisions from one designer must be available for other designers as quick as possible, because design decisions possibly influence the decisions of other designers.
* to shorten the period of time for the design and also for the manufacturing of a product an integral part of a new CAD system must be the use of standard parts and company specific standards. So, there should be the possibility to use a centralised library with standardised parts. In addition to this library there is a need to have the possibility to define which standard parts can be used for a certain product or project.

4 Modelling the Concurrent Design Process

4.1 The Design Process

Hubka [8] defines the term design methodology as: 'General theory of the procedures for the solving of design problems. It concerns both the strategy of proceeding, i.e. the general path, and also the tactics of action in small portions of the work'. Preliminary attempts to define a general design methodology were developed mainly by university professors who had learned the art of design in practical contact with products of increasing complexity. Mr. G. Pahl and W. Beitz enumerate in their book 'Engineering Design' [13] a list of different approaches to modern design methods according Hansen [6,7], Rodenacker [17], Roth [19,20] and Koller [10,11]. The different approaches were integrated in one general approach which is described in [22].

Stages of the Design Process
The design process can be defined as 'Work flow of design, leading from the problem formulation to the complete description of the technical system' [8]. In general each design process may be structured into four main stages, that can in turn be divided into a smaller or larger number of subsidiary activities - design steps [13] (figure 14): Clarification of the task, conceptual design, embodiment design, and detail design.

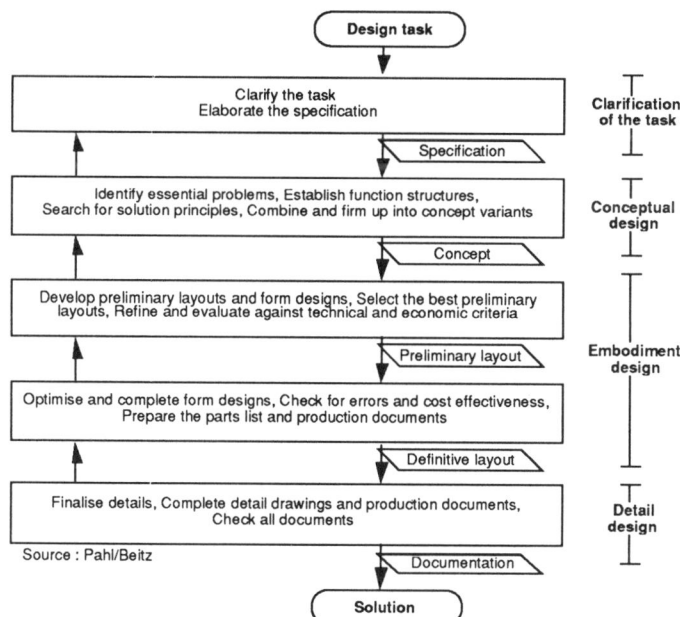

Figure 14: Stages of the Design Process

The procedures during design, and the appropriate working methods, are contained in general form in design methodology. During the design process several important documents or provisional results are created:

The *specification* is the requirements list of the customer. It describes all relevant aspects and demands without whose fulfilment the solution of the design task is not acceptable. Additionally there could be listed wishes that should be taken into consideration whenever possible. The specification also includes information about quantity and quality.

The *concept* describes the solution of essential problems of the design task by function structures and solution principles. The basic solution path is laid down trough the elaboration of a solution concept.

The *preliminary layout* describes the embodiment-determining main function carriers; that is the general arrangement, component shapes and material.

The *definitive layout* describes the overall layout design (general arrangement and spatial compatibility), the preliminary form design (component shapes and materials) and the production procedure.

The *final documents* of the design process are detailed component drawings, assembly drawings and parts lists. Depending on the company there are other documents like quality control documentation, operation plans and so on.

Description of Design Tasks

A design task can be described by the function of the needed solution and a set of requirements and constraints. The *product function* can be used as an abstract and solution independent description of a design task. According [13] the product function (or overall function) can be decomposed into several sub-functions. Figure 15 presents a functional decomposition of a stamping and patching unit.

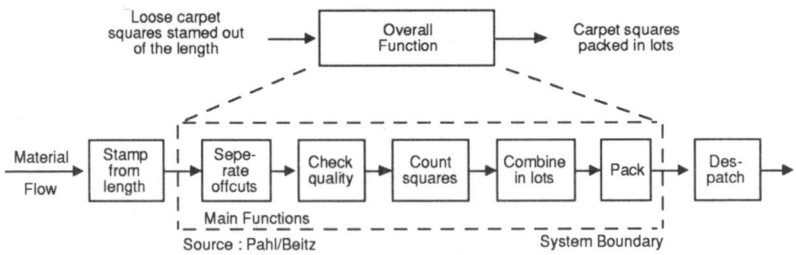

Figure 15: Decomposition of Product Function into Sub-Functions

The relationship between sub-functions and the overall function is governed by certain constraints, in as much as some sub-functions have to be satisfied before others. The interrelationship of sub-functions produces a so-called *function structure*. It is useful to distinguish between main and auxiliary functions (figure 16). While main functions are those sub-functions that serve the overall function directly, auxiliary functions are those that contribute to it directly. They have a supportive or complementary character and are often determined by the nature of the solution.

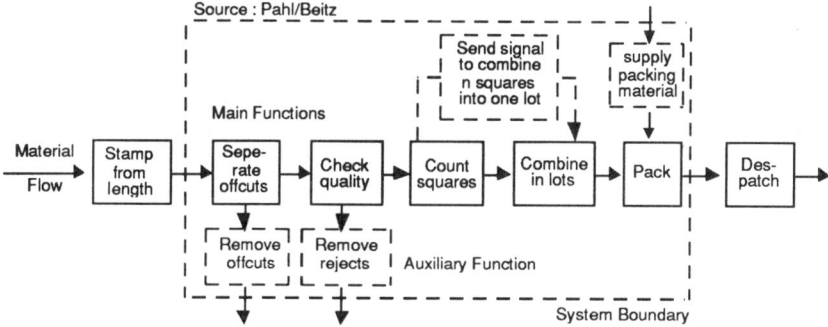

Figure 16: Decomposition of Product Function into Sub-Functions

Establishing a function structure facilitates the discovery of solutions because it simplifies the general search for them and also because solutions to sub-functions can be elaborated separately. Sub-functions are usually fulfilled by physical (or chemical, biological) processes or phenomena (figure 17).

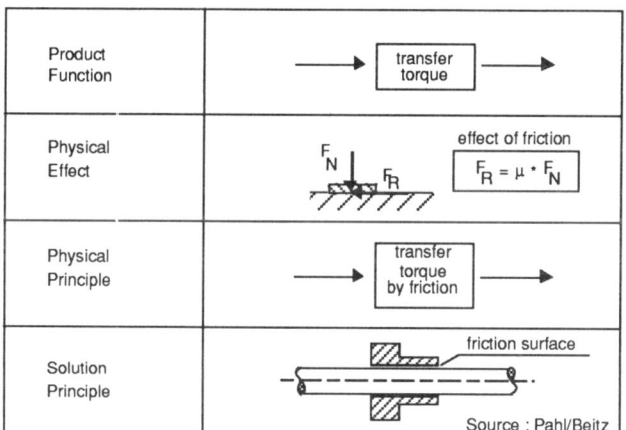

Figure 17: General Search of Solution Principles

Physical processes are based on *physical effects* like the effect of friction. Several physical principles may have to be combined in order to fulfil a sub-function. If the effects are assigned to sub-functions, we obtain the *physical principle* of that sub-function. To the end, the required layout and preliminary forms have to be specified to obtain the *solution principle*.

4.2 Design Project Management

The presented design methodology is very general and does not say anything about *how* the design process is organised. It enumerates the general stages and methods to solve design tasks but do not solve team work problems.

The comparison of the presented design methodology and the design process found at the company Dessindus shows that there is no difference in the order of design phases or its contents. In addition to the presented design process Dessindus makes an additional distinction in the embodiment design phase: the preliminary design and the general design phase.

An important result from the analysis of concurrent design process at Dessindus is that concurrent design normally starts after the preliminary design phase, which is part of the embodiment design. The work done in the earlier stages of the design process can also be done by several designers but generally this people have mostly the same work to do and try to find the best solution for one problem (i.e. brain-storming methods).

The reason for starting concurrent design after preliminary design seems to be the possibility to define design sub-tasks in an accurate form. If the definition of the sub-tasks is not exact enough the mistakes and expense of communication and control are larger than the efforts of concurrent design. It is also important to notify that the sub-tasks done simultaneously should have a certain complexity.

4.2.1 The Project and the Product Model

From the knowledge that within a concurrent design process not only product data is important but also process and organisation data, within the CACID project a main distinction is made between project and product data. Project data should describe the organisation oriented view and product data should describe the product oriented view of the design process. This leads to two main interrelated models: the *project model* and the *product model* (figure 18).

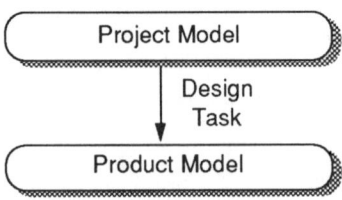

Figure 18: Project and Product Model. Two Views of the Design Process.

4.2.2 General Project Data

The production schedule of a company results in the creation of developing projects. This means that projects, designers, design tasks and the produced product descriptions have to be managed. A simple model of this object types is given in figure 19. As a description method the Object Modelling Technique from James Rumbaugh [21] ('Object-Oriented Modelling and Design') is used which includes a good graphical representation for object-oriented models. Object types are symbolised by squares and relationships between them are symbolised by lines with different attributes to describe their meaning and cardinality.

Each design project is specified by a project name, an acronym, a description and other company specific data like a starting and an ending point. There may be further attributes attached to a design project depending on customer needs.

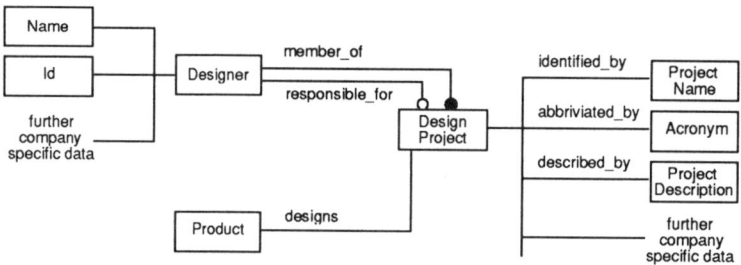

Figure 19: Description of the Object Design Project

All designers of a company are listed in an appropriate agenda which can be modified by adding or deleting designers. Designers can be assigned to a project as a member of the design team. One of these designers is the responsible designer of the project. The designer model can also be used to store important time planning data like which designer is available on which days of the week or if a designer is on holiday and can not react on demands from other designers.

As already mentioned it is important for the concurrent work of several designers to have the ability to exchange explicitly information among them. For this purpose a special mail tool - the bulletin board - is introduced. With the aid of this tool mails can be send from one designer to one or more designers (as well as to all designers of the company). Mails can also be send to design projects, this means that the mail is distributed to all designers of one design team (project news). If the design system itself recognises changes which are important to certain designers or projects it should have the possibility to send mails (figure 20).

There are two different types of mail. The first one is the 'problem'-mail which can be used to announce important design problems which must be solved as quick as possible. The second one is the 'info'-mail which can be used to distribute common news among the designers.

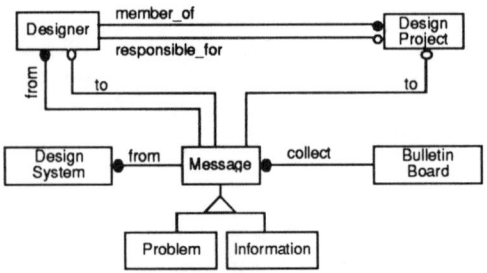

Figure 20: Bulletin Board Facility for Communication

4.2.3 Design Task Management

The concurrent work of several designers on the same product requires, that the overall design task can be decomposed into sub-tasks - depending on the aspired technical solution - which can be developed separately by different designers. A design task management system should allow the creation and description of design tasks and the visualisation of the resulting hierarchical decomposition structure of the overall design task. A proposed dialogue window is shown in figure 21. The used icons are explained in figure 22.

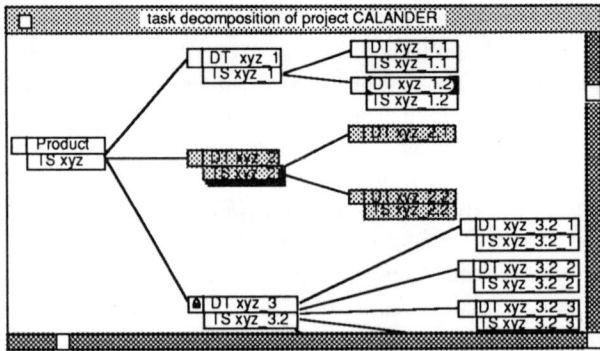

Figure 21: Design Task Decomposition Editor

For each design project, there exists one overall design task, which is shown on the left side of the decomposition board. To each design task one or more technical solution (TS) can be attached. This technical solution appears beneath the name of the design task (DT). Each technical solution can be decomposed into several other design sub-tasks. If there exists more than one technical solution to a design task, this is notified via the shadow box behind the technical solution box.

The task decomposition board should give information about the current status of the sub-tasks as well. If a designer is just working at a part of the product model, the design task and its technical solution are displayed inverted. This means except this particular designer, nobody has the right to make changes to this part of the decomposition board.

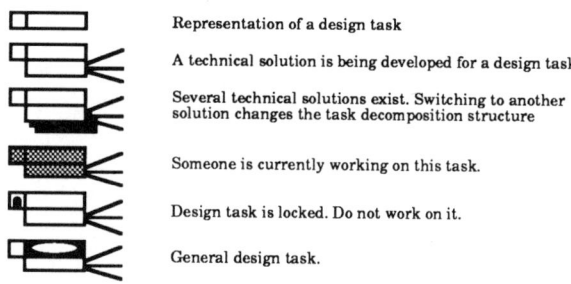

Figure 22: Symbols used in the Design Task Decomposition Editor

For the description of the contents of a design task additional dialogue sheets are needed, i.e. for the description of special types of functional requirements - like standard parts requirements.

If a designer starts solving a design task he creates a technical solution. There may exist null (at the beginning), one or more technical solutions for one design task. Depending on the aspired technical solution the design task decomposition is different for each technical solution.

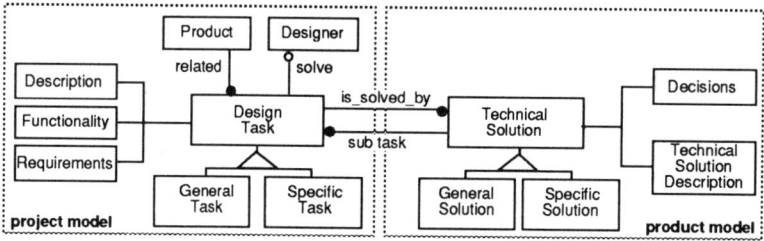

Figure 23: Description of the Design Task Decomposition

A simple model of the task decomposition structure is described in figure 23. A task can be described by several attributes. Attributes may be i.e.. the product functionality, the design task description and requirements. Requirements can either consist of constraints concerning available space, standard parts or other more general requirements. Some of the attributes are company specific.

Two general types of design tasks are proposed: the general and the specific design task. A specific design task depends on other parts of the overall technical solution. It is a task with the aim to change a certain area of the product. The location and the context of the aimed technical solution is clearly defined. A general design task is a task which can be solved separately. Later on its technical solution (it is a general technical solution) is incorporated (possibly on several locations) into the product model and solves one or more context specific design tasks.

A technical solution is the representation of the results of exactly one design task. If a technical solution of a design task is to complex to be solved by one designer, it can be divided into several new design sub-tasks and so on. This decomposition results in a hierarchical structure of design tasks. A technical solution is described by a list of important technical decisions that have led to this particular technical solution and the solution description itself. Technical decisions can be divided into internal and external ones, i.e. decisions that have a strong influence on other design tasks should be marked external, so that other designers recognise the decisions which influences their work. Internal technical decisions are only for documentation purposes.

4.2.4 Managing Technical Solutions

As described before, technical solutions should be worked out when design tasks are defined. All technical solutions together should define a consistent product model which fit the requirements of the customer. So, each technical solution must fit to the rest of the product model. But how to synchronise the different design activities and the resulting technical solutions?

A design activity creates or modifies product data. These modifications may be small or large and may need seconds, hours or weeks. In the following this is called a *design transaction*. Such a transaction transfers a *product model state* into a new state (figure 24).

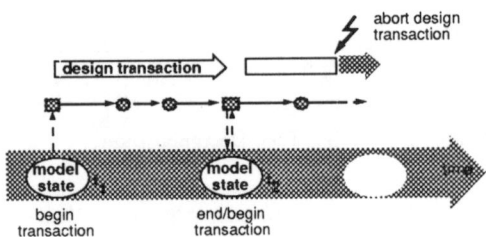

Figure 24: Sequential Design Activities

Such a transaction can be aborted on demand and the last consistent product model state remains unchanged. If several transactions are running in parallel (figure 25) - which is the goal of concurrent design - then probably inconsistent product model states may be created.

To avoid this, two basic methods are proposed. One method is the definition of *design spaces* before a design transaction starts (a design space is a task and requirement definition that helps to avoid design conflicts by clear task distributions and descriptions). The second method allows every designer to create his own product model states which then behave like *product model versions* and *alternatives*. To bring different product model versions together is the task of a merge procedure.

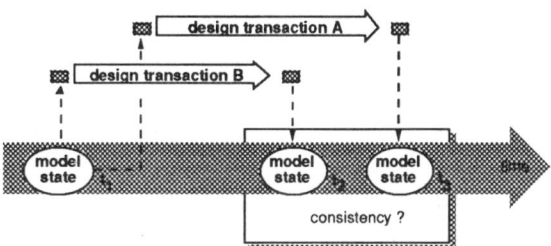

Figure 25: Parallel Design Activities

When we look at the product development process as a continuously improvement of the product model, then the synchronisation problem is to get a consistent sequence of product model states, even if several design activities are modifying this model simultaneously. From the requirements of the product development process this results into three points of time, where synchronisation should take place:

1. Before a design transaction starts, to distribute tasks and mark off the task descriptions.
2. After a design transaction has finished, to correct conflicts and to harmonise partial results.
3. During a design transaction, to exchange ideas and discuss interim results between designers.

The solution worked out in the CACID project are based on the idea that product development means creation of product model versions (one product model state is a product model version) and alternatives. Each design transaction creates a new product model version. When several transactions / designers are working in parallel they create product model alternatives which must later be harmonised and result in a new product model version (figure 26). These alternatives are located within the *workspace* of a designer. In such a workspace a designer may again create versions and alternatives of the product model to solve his design task. Nobody else should have the possibility to look in this workspace without the permission of the responsible designer.

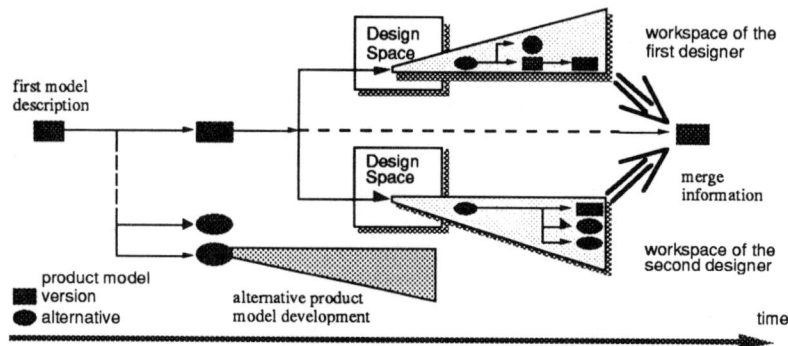

Figure 26: Using Product Model Version in a Concurrent Environment

Before an activity may create a product model alternative a design space must be created where the task, its requirements and constraints are defined. By comparing the design spaces of active design transactions it is possible to synchronise them before they start. Synchronisation after an designer's transaction means that different product model versions must be merged into one harmonised result. This can be done partially automatically by comparing new, modified and deleted information. This merge process also may use the design space information to decide which transaction was responsible for which modifications.

To support synchronisation during a design transaction simple communication facilities may be used. One possibility is to give the permission to look at the intermediate results within other workspaces or running transactions. Another possibility is to exchange product information via a clip-board or an electronic mail facility and integrate this information into the running design transaction.

The development within product model versions by single designers is accompanied by the description of the project data. Here the reasons for developing a version and the responsible designers are specified.

Both models together (figure 27) give together an exact description of the state of the concurrent design process. The product model version graph presents all product information and its history. Especially alternative ideas are documented. The task decomposition structure gives a good overview about design tasks, its states and describes links to the worked out technical solutions.

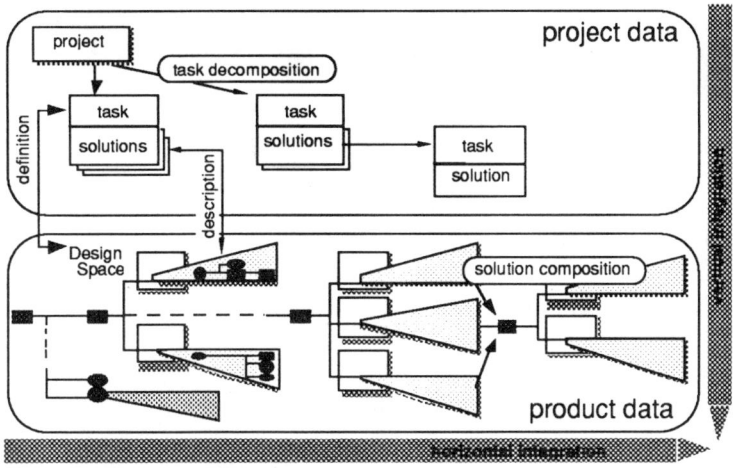

Figure 27: Integration of Project and Product Data

4.3 Description of Design Data

Design data is all data that is generated during the development of a product and which is used to produce it. This includes everything from market surveys to workshop layout. A goal of the CACID system is the support of the design process as a whole and hence all data that is used during process development must be managed in the system.

Decisions and information usually flow from general conceptual design to more and more detailed specifications of the product. To find an optimal solution the design must be reviewed at any stage. This may result in updates of the information in earlier stages. The design data must be described in a way to make this possible.

Until now design data is not necessarily stored on a computer. It may also include conventional paper documents. However in order to support all design activities all documents should also be an integral part of the data model. In general though, design data is edited and managed in electronic form. This allows efficient management, access and update of all information. In the CACID project the central notion that describes design data is the technical object. It is used to organise design data and to access it.

Design data does not only concern the information as such but also the way it is organised. There exist links from one piece of information to another. These links represent the structure of the data and they are just as important as the contents themselves. In order to achieve simultaneous engineering the links between the data elements must be kept alive and used as a basis for integration. During the product development process people from different departments work on a design. Each department is responsible for a different aspect of the overall development. They are all adding their own special expertise and information to the product.

In each of the design phases specific information is worked out. In a traditional setting all the information is captured in separate documents. Documents are then transformed from early stages to later stages through to the completed document. However the goal of the CACID system is to go beyond document management. Wherever semantic models of an application domain are available CACID models the data to provide identical means of accessing and maintaining design information for all people involved.

4.3.1 Integration of Design Data in a Product Data Model

Product modelling tries to integrate the knowledge from all the different sources, to make them more easily available and to keep them consistent. Integration can either be at the level of the tools or the data itself can be integrated. In the first case the tools operating on subsets of the data model are integrated in a framework. In such a framework common functions are provided for the tools. However the data models of the tools remain separate and only interface with the use of post processors. This produces problems of data redundancy, loss of information and slow access.

In an integrated product data management system all this information is unified in a single data model. Hence a tighter integration is only possible in a single integrated product data model that supports all aspects of product development. Such a general model is worked out in the STEP activities. The slow progress of this project shows that the problems encountered are very complex and that the model is hard to implement. Some parts of the product data model can not yet be integrated.

CACID does not attempt to provide a complete implementation of product modelling, but rather stores it's non geometric product data in the fashion of a product model. The technical object model integrates the partial models wherever this is feasible, but keeps incompatible models separate until they can be handled. This improves the integration as compared two framework systems and is able to deliver results earlier than the fully integrated approach (figure 28).

In the development of the technical objects model an evolutionary approach has been taken. Subsets of the product data model were identified which are able to extend current CAD/CAM technology in short terms. This product model can then be extended step by step to incorporate more data and to provide more sophisticated algorithms on the product model. As more tools for specific sub models become available, these can be integrated and as a result the technical object model will gradually evolve into a fully integrated product data model.

This means, that new functionality developed must be implemented as an extension of existing functionality. Otherwise the information flow through the system will be interrupted, and data has to be supplied by hand without need.

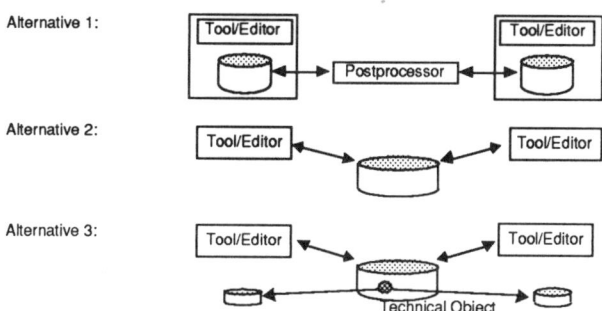

Figure 28: Alternative Integration Approaches

Semantics of Product Data Models

A data model is an abstraction of real world facts. Such an abstraction is necessarily limited in many cases. Frequently a model only captures the syntactic structure of the domain it models, but no tools are available, that ensure semantic integrity. I.e. a 2-dimensional drawing only looks like a real world object if the designer makes no mistake, but it is not possible, to check whether an object drawn can actually exist.

For some sub models however there are algorithms that can understand the semantics of the data. These areas are the most interesting, because the computer has complete information. Examples include 3D geometry, structural information, and mathematical models.

In areas, where tools, ensuring semantic consistency, are not yet available design data can be represented on a computer as text or as pictures which are not interpreted, but only stored for common access and modifications. For these documents, the user can specify the links to other information and store them in a way that allows him to control access and updates.

Views of the Product Model

Each person involved in product development has its own view of the design information. This view is usually a subset of the complete information mapped into a specific domain. Each domain is supported through special tools tailored for that domain. A domain may be drafting, NC-programming or marketing.

From a complete product model these views can be generated and accessed by the designers. However starting with an unspecified product it is very hard to generate the product data model from the views. CACID takes the approach of providing views where they can be generated. Otherwise the different views are stored in the common structure of the technical objects model, allowing easy access.

From Documents to a Product Model

When design documents are analysed in order to integrate them in a product data model the following questions arise:

- What is the type of data contained in the document?
- What purpose does it serve?
- How is it structured?
- How is it linked with other documents?
- How is it updated?
- How can it be incorporated into an integrated model?
- Is there a semantic model for it?

Another important aspect of design data in an integrated product model is how the maintenance of the information can be automated. Only data that can be generated and maintained automatically can be guarantied to be always consistent. If this is to be achieved the complete semantics of the data is to be known. Then the computer can interpret the information and work on it directly. Today this is only possible in some areas of product development. Currently mainly 3D geometry can be maintained consistently, other areas like requirements engineering or technical computations have reached different levels of modelling.

4.3.2 Contents of the Product Model

In order to build up the product model a structure is required which provides a skeleton where information from the various sources can be attached to. Hence we need to work out a description of the particularities of each design view, and an overall mechanism to integrate them.

Design stage	Integral design data	Concurrent design data
Clarification of the task	Requirements engineering	project / team organisation
Conceptual design	Functional structure, technical objects	Task decomposition
Preliminary design	Constraints, overall dimensions	Creation and management of design spaces
Embodiment design	Technical objects mapped to parts or standard parts	Alignment and interconnections of components
Detail design	Attribute values	Geometry

Figure 29: Main Types of Design Data

For each stage of the design specific types of data are modelled. Each kind of data needs a different editor. Figure 29 shows what kind of data is required for a specific stage of the design process: We can see that in the column of concurrent design we are mainly involved with geometric data, whereas in the column of integral design we are mainly concerned with structural information and data in the form of values/formulas/text. The CACID system has to make sure that all kinds of data can be edited.

Basis of the Product Model

The underlying structure of the product model is usually derived from one type of data, that is considered most important. The structure of this data provides the organisation used to attach the non geometric information to.

In conventional systems, which try to incorporate product information this is either the part/assembly structure or sometimes the geometric structure (form features). This has the disadvantage, that these types are usually available late in the design process or not available at all. Also the product information has a different scope than the geometric information or a bill of material.

In a simultaneous engineering framework this must be a structure that can be worked out early and that is usable in all stages of product development. It was found that during product development so many different kinds of evaluations (as costs, quality analysis, production planning, etc.) are based on this information, that it was appropriate to develop a specific model for it.

To find a basis for the structure of the product model the engineering design process can be followed as a paradigm. The task that occurs first during design is the functional decomposition of the product. This decomposition can be made very early, without reference to geometry or other data. It also provides a efficient mechanism to distribute the tasks to the designers who work in parallel. Functional units have a high degree of coherence, which allows independent development.

Each functional unit in the design is represented by a technical object. Technical objects are grouped hierarchically according to sub functions.

Structure of Design Data

The key point in this model is the creation of structure. Structural information is used to decompose and group the information in the product data model. The links that specify references from one object to the other create the structure. In the simplest case, objects are organised in a hierarchy. However data modelling frequently requires references, that can only be specified by graphs or with cyclic data structures. Even if nothing is known about the contents of a document, it can be stored in the right place in the hierarchy. Using this structure efficient access and update control can be realised. The user follows the links from one technical object to the other to gather the information contained within.

Technical Objects

A technical object is an abstract design entity which is represented by a node in the product data structure. A technical object is the aggregate class of all representations of the unit. It consists of links to other referenced technical objects and of a set of attributes (figure 30). The links may have different roles, such as 'sub function', 'part_of', 'attached_to' etc.

The designer can create arbitrary technical objects as nodes in the structure and attach data to it. Since a technical object can also be a piece-part, the technical object structure also contains the assembly structure as a subset.

A technical object can have several representations. A representation is a description of the technical object in some form. Each representation can in itself be a structured document like a CAD drawing, a 3D model, or a set of engineering calculations. Each representation is independent of the other. Their contents are all gradually developed as the user inputs data into the system.

For each representation there can be several presentations. Presentations are mappings of the data that can be performed automatically without loss of information. Examples are the presentation of a body in wire frame form or as a shaded picture. Other presentations are symbols or simplified rough shapes for certain parts.

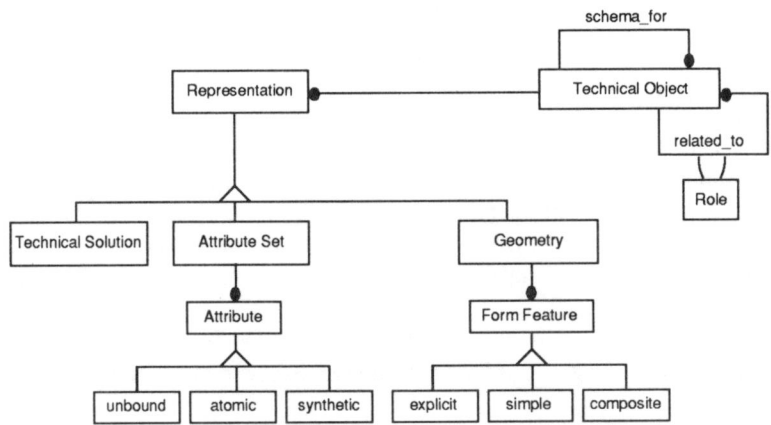

Figure 30: Technical Object Data Structure

Since technical objects are independent of the representations, the data in the representations can be changed arbitrarily without loss of the structural information. The technical object always remains identical even if its representations change or are deleted altogether.

Attributes
Each technical object maintains a set of attributes. An attributes is a piece of information with a name and a value. The attribute name can be used to reference the attribute from other technical objects. Synthetic attributes can also be computed from other attributes through formulas. On the basis of formulas automatic computations can be included in the technical object structure.

Geometric Design Data

The most important class of technical object representations are geometric designs. They are the output of the CAD systems employed to develop the product.

The underlying geometric model of CACID is a form feature representation. Form features are used to generate the 3D boundary representation model from a parametric constructive solid geometry tree. The parameters of the form features are accessible as attributes from the technical objects.

4.4 The Need for Standard Parts and their Description

To be competitive for small manufacturers it is necessary to use as many standard parts as possible. This shortens the development and the manufacturing period. In addition, the use of standard parts eases the maintenance of the products and improves their quality.

The standard parts sub-system should be divided into two components:

- the actual standard parts sub-system, which provides the services needed for the generation of standard parts inside the CACID system, e.g. the communication with the CACID kernel,
- the standard parts sub-system library, which provides the actual parts, their geometrical representations, the standardised parts list information, etc. Such libraries can be created either by standardisation committees, part vendors, etc.

The standard parts sub-system has the goal to generate pre-defined parametric parts using parameter tables. These parts are stored as technical objects in the CACID product models. The sub-system implements the concepts of:

- parametric parts,
- standard part hierarchies, and
- attribute dictionaries.

As a technical object, a standard part can be defined as a principal element, a rough shape model, or a detailed model. It can appear in parts lists and therefore should have identifying names suitable for such lists. As a technical object, a standard part must provide data for position and orientation: reference points, vectors, connecting points, local co-ordinate systems, and 'fit in' or 'fill in' relationships. Additional information such as: tolerance, material, surface, physical function, could be provided in the standard part description.

A standard part can be established by constructive geometry. This constructive geometry is realised through an interface by using the corresponding CAD geometrical entities.

As a consequence for the standard parts sub-system there is a need for attributes, several geometrical representations, structures, and access methods to standard parts data.

Standard parts attributes:

- a standard part must refer to the definition level: principle element, rough shape or detailed model, and
- a standard part must have identifying attributes to be displayed on a part list.

Geometrical representation:

- geometric primitives are to be used in the geometry interface,
- a form feature can be positive (protrusion feature) or negative (depression feature),
- positioning and orientation information must be provided, and
- 2D or 3D representations must be considered.

Structure of the standard parts sub-system:

- relationships between principle elements, rough shapes, and detailed model must be implemented, and
- multi-suppliers catalogues must be considered.

To realise the library, the CAD data model must be enhanced by the following data types:

Dictionary:	class, class hierarchy, view, view hierarchy, attribute, attribute type
External representation:	external name, multilingualism, short name, long name
Additional information:	definition, help, note, example, icon
Internal representation:	class attributes, view attributes, instanciation method, control method, realisation method

The CACID standard parts sub-system solution is upward compatible with the existing DIN V66304 solution, but also introduces an object oriented extension, similar to the CEN/CENELEC approach. The following documents have been used:

- CADLIB Logical description Format of supplier library,
- CADLIB Representation Transmission Interface,
- CADLIB Conceptual Model,
- DIN V 66304, and
- DIN Technical Report No. 14.

The standardisation on the International Standardisation Organisation (ISO) and Comité Européen de Normalisation (CEN) level was going on rapidly, so the object was to follow the movements to use a snapshot as basis of the CACID experimental work.

The conceptional model in the CADLIB project differentiates three standard parts library users (figure 31):

- Library suppliers, who describe a set of standard parts using a standard format. They guarantee for the quality of the data.
- Library integrators, who provide a software environment to integrate the supplied library.
- Library (end) users, who use the supplied and integrated library.

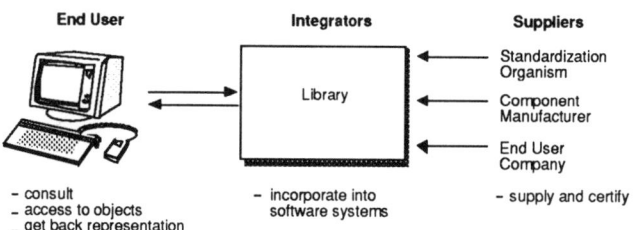

Figure 31: Standard Parts Library Users

The general overview of the standard part library is presented in figure 32. The main concept is that an object can be viewed in several ways, such as a principle element or as a detailed element. The dictionary is composed of object entries and view entries. The external representation allows views of these entries that can be understood by the user. The internal representation allows views that can be realised by the CACID system.

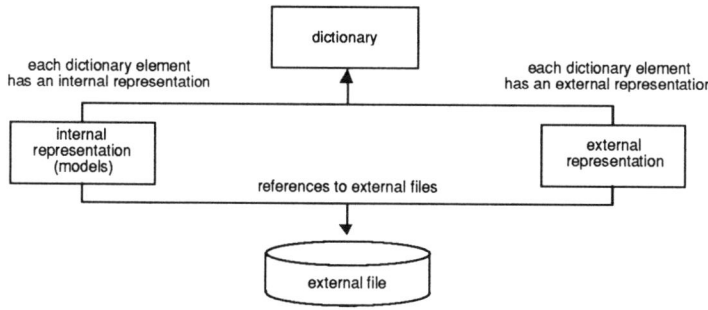

Figure 32: Conceptual Level: General Overview of a Library

The programming language used for the description of a library is C++, which is used also in the CACID project. Such a language, with its object oriented capabilities is very useful for the description of a library. The basic point of this implementation is the model. A model is a description of an object for the realisation of a view. Inheritance mechanism of C++ are used. For each model at least the following methods have to be defined: get_attribute_values, control_attribute_values, realise_view.

5   Implementation of Computer Aided Concurrent Design

The concepts described in chapter four leads to the implementation of the so-called CACID system. It should support as well product modelling as concurrent design functionality. The system was developed in several mayor steps which are described in this chapter.

5.1 Implementation Idea

From the designers point of view the CACID system (figure 33) is a concurrent CAD system with advanced features in the area of:

- product modelling,
- acquisition and usage of standard parts, and
- concurrent design support.

Its data storage manages product data, design project data and standard parts data. The standard parts library is supported by a data acquisition tool which allows a comprehensive data management and input.

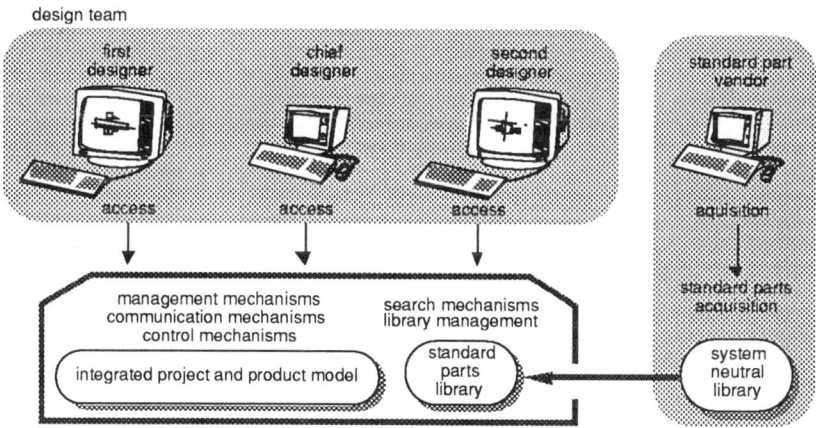

Figure 33: The CACID System from the Users Point of View

On the implementation level the CACID system consists of a set of tools like a conferencing and communication tool, a design project management, a product modelling tool, a standard parts acquisition tool, a standard parts library and various data management tools. Most of the tools were realised within the project as prototypes. For some of them only concepts exist.

5.2 System Architecture

At the beginning of the CACID project an early prototype system was realised to test various aspects of the conception ideas described in chapter four. In July 1992 the project consortium decided to use the results of the early prototype in two ways. One approach was to refine the existing CACID prototype to test the possibilities of design space definition, reservation, release and merge procedures. The second approach was to use part of the prototype (the design project management component which handles the project information, product model versions, design space reservation information etc.) and link it with the SOLID 2.0 modeler which is a commercial 3D CAD system based on the geometry kernel ACIS from Spatial Technology developed by the company Strässle. The main reason for this decision was that it was not possible within the short duration of the project to develop a complete product modeler with all the interaction procedures and the user interface needed to make it usable for validation in a real design office.

5.2.1 Realised First Prototype

At the beginning of the project the early CACID prototype was developed to answer central questions like:

- usage of the geometry modeler ACIS as the kernel of a product modeler,
- usage and integration of the object oriented database system Objectstore,
- integration of project and product data.

The User Interface and Architecture of the CACID Prototype
The user interface of the prototype (figure 34) consists of an icon area and several graphic windows. The icon menus are used to specify commands in a very quick manner. One pick into the icon area often replaces a complex set of operations. The top icon area is used to start the most important commands. With the help of some buttons sub-menus can be swapped and problem specific actions can be activated. The graphic windows are used to display shape information of the designed product. By choosing a view point different views can be created.

In addition to the described user interface objects a set of so-called 'editors' and dialogue sheets can be opened to edit i.e. project information or to describe design task decompositions (see also figure 38,39,40).

The prototype is completely object oriented designed and implemented in the programming language C++. The architecture of one CACID workstation can be decomposed into four main system components (figure 35):

- The user interface,
- The product model,

- The project model and
- The object-oriented database management system.

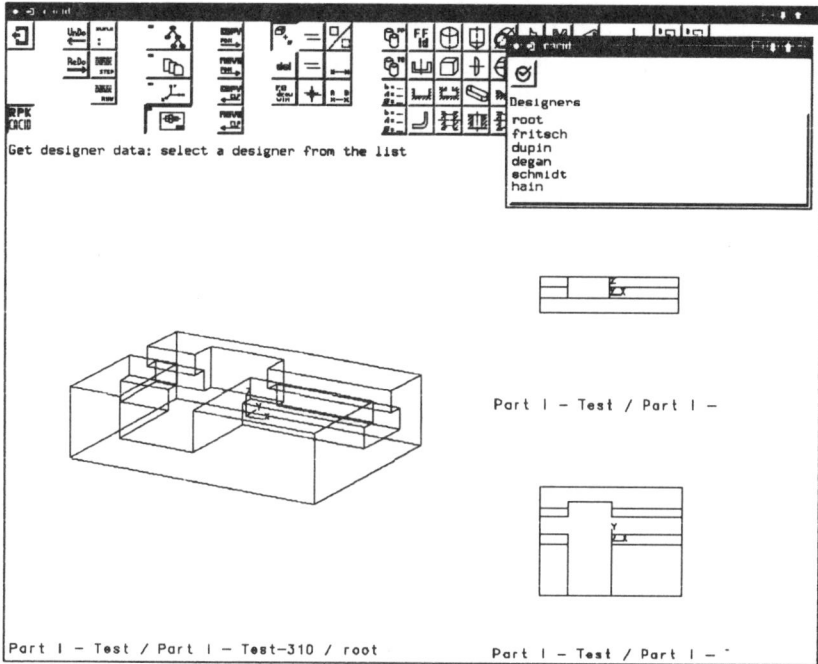

Figure 34: User Interface of the First CACID Prototype

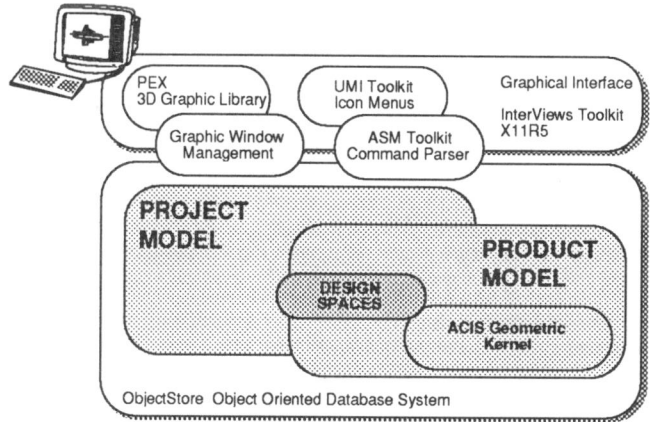

Figure 35: Architecture of a single CACID Workstation

Some of the components are completely developed by the project, other components are based on commercial available software libraries.

The user interface component:
The user interface consists of several sub-components. As the basic graphic library the most recent X-Windows System release X11R5 is used. In addition the C++ user interface class library InterViews is used. This public domain library includes a set of predefined window objects like buttons, text-widgets, etc., which hides the X-Window System programming interface.

The menu and ASM tool kit are used to support high-level programming of user interactions. The menu tool kit allows the specification of icons, windows and so called direct actions. A direct action is an action which immediately starts after a click into a menu icon. This means that no further input is needed. The ASM tool kit is used to support the programming of complex user interactions which need different kinds of input and can include several input alternatives and various command passes.

The 3D graphic extension of the X-Window protocol (PEX) is used to convert the 3D geometric representation of the product geometry into a 2D view on the screen. The graphic window management component is used to manage several graphic windows. It controls the relationship between a product model and a view, stores the chosen view point and manages the list of visible objects within every view.

The project model component:
The project model includes all objects needed to describe and manipulate the main design project information. This is i.e. the list of the defined design projects, the list of available designers, design task description and decomposition.

The product model component:
The product model includes all objects describing and manipulating a product definition. One sub-component of the product model is the commercial software package ACIS. This sub-component includes all objects needed to describe shape aspects within a boundary representation structure. In addition the product model includes all data and methods needed to describe form features and design spaces.

The database management system:
The database management system is used to store project and product model data in a persistent way. The used software Objectstore, developed by the company Object Design, supports a C++ programming interface and can be used as a distributed database. Within the CACID project the database is used to create persistent data after every database transaction. This means, that every change within the project or product model is directly stored into the database. This is used to support simultaneous access to common data by several designers.

The distributed CACID system consists of several workstations and the CACID server (figure 36) which may be a single workstation or part of a CACID workstation. All workstations are connected with a local area network.

Within the distributed system only one project and product model exist. It is stored and managed by the CACID server which provides something like a virtual persistent memory which can be accessed by every workstation. Virtual the data is available on all workstations.

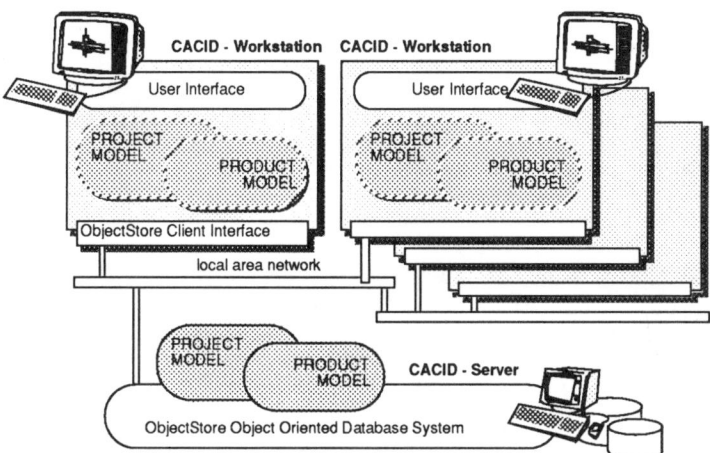

Figure 36: Architecture of the Distributed CACID System

As a result from this architecture i.e. every change of a product model is visible on all other CACID workstations. But because every designer works on his own product model version this does not result in any direct access conflicts.

Handling Parameters in the CACID Model

The dynamic and user defined definition of parameters play an important role within the CACID system. By example the project model includes a set of objects (design projects, designers) which should be customisable by the company who uses the system. An other example for the usage of dynamic defined parameters are form feature descriptions with technical attributes in the product model.

To support object-types with a dynamic defined set of parameters the CACID model includes a small sub-model (figure 37) with the classes Parametric Entity and Parametric Value Entity. Every class which is derived from this object(s) may define a variable set of parameters and store the related values.

Figure 38 presents a dialogue sheet of the CACID prototype which allows to fill in project data. The initial parameter list definition is show in the second window. According the type of the parameter (string, text, integer-number, real-number) the dialogue sheets are configured at run-time.

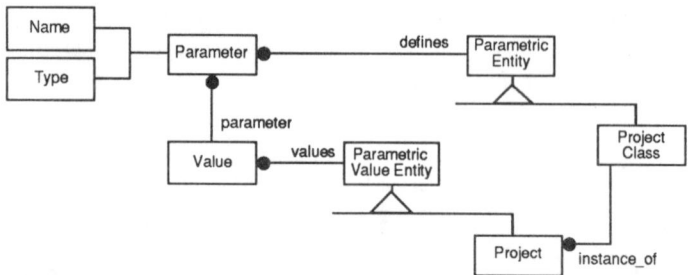

Figure 37: Simple Parametric Model of the CACID System

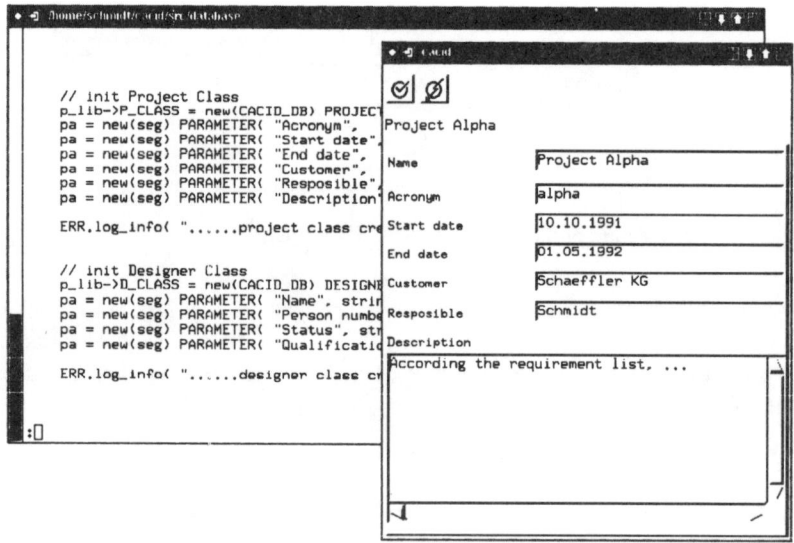

Figure 38: Project Dialogue Sheet and its Parametric Description

Design Task Decomposition Editor

A design project complies the task to design a new product. A design task can be solved by one or several different technical solutions. For each aspired technical solution it may be necessary to solve a set of new design tasks. The initial given design task can therefore be split up into a hierarchy of design tasks and aspired technical solutions. Figure 39 shows the graphic editor of the CACID system which supports the description of the design task decomposition.

The editor consists of a command icon area and a graphic area where the design task decomposition is displayed. A design task or technical solution is symbolised by a square with a short name and the name of the responsible designer. A pick into one of the symbols opens further dialogue sheets to read and edit more information. If a design task has several solutions a set of solution symbols is displayed. By picking a solution which is displayed in the background the actual technical solution - design task decomposition is replaced by the structure of the picked solution. Technical solutions which are currently in work are displayed with a red colour.

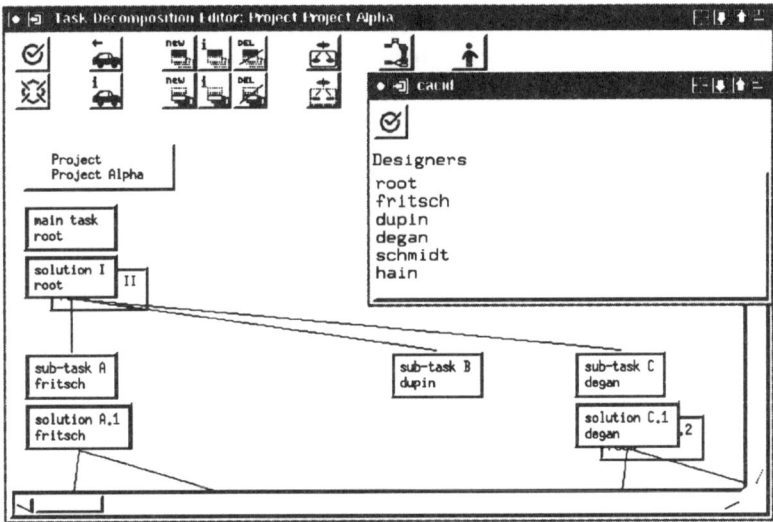

Figure 39: Design Task Decomposition Editor

The described editor and other dialogue sheets not presented here support the creation of new projects, the modification of project information, the definition of new parameters for all projects, the attachment of designers to projects and design tasks, the decomposition of design tasks and the textual definition of requirements to technical solutions.

The Product Model Version Graph Editor

As described in chapter four technical solutions are part of a product model. If a designer works on a technical solution the corresponding product model - where the solution is included - is modified. A sequence of such design steps within a design task is called a *design transaction*. A design transaction may last minutes, hours, days or even weeks. Before a design transaction can be started a new product model version is created. All modifications during the design transaction are done within this model version. Only one designer can work on a version at a time. If another designer starts another design transaction in parallel he works on a product model alternative.

As a result from the design transaction concept and the creation of versions and alternatives the designers build up a version graph which represents the overall modifications of the product model during a design project. The version graph editor (figure 40) lets the designer inspect the product model versions and alternatives.

If a designer starts his work on a design task he must choose a product model version - probably the latest version - where he likes to work in his modifications. If the design transaction is finished the version can be made frozen. This shows other designers that this version can be used for further design tasks. To bring together different product model versions which were worked out in parallel a merge process is needed. Within a merge task a designer takes one product model versions, works in all changes of another version and then stores the new - merged - version.

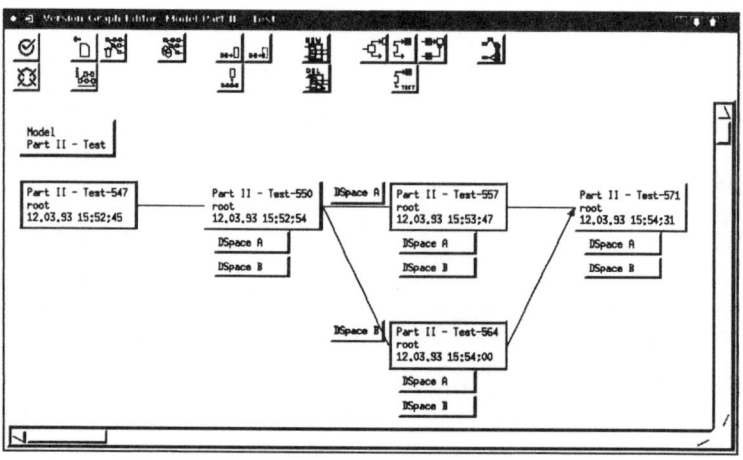

Figure 40: Product Model Version Graph Editor

The merge process may be simple or complicated. This depends on the quality of the co-ordination between the different design transactions. In general there are three points in time where a co-ordination between two design transaction can take place:

- Before several parallel design transactions start,
- During parallel design transactions are running,
- At the end of parallel design transactions.

All three possibilities are included in the CACID prototype. The second and third possibility are clear. Co-ordination during design transactions means to have the possibility to communicate with other designers or simply to have the possibility to inspect the intermediate results of a running design transaction. Co-ordination at the end of parallel design transactions this is realised by the merge process.

Co-ordination before several design transactions starts means to assure that certain design constraints or requirements are respected by each design transaction. As more exact these constraints are, as more easy is the merge process at the end of parallel transactions.

The definition of design constraints is done in the CACID system with the help of so-called design spaces. The formal definition of design spaces depends on the available data in the product model. If the product model only includes shape information then only this type of information can be constrained. Naturally in addition non-formal information like textual descriptions of design constraints can be part of a design space definition.

In the following two sub-chapters the contents of the product model realised in the CACID prototype and the usage and definition of design spaces is described.

The Geometric Kernel ACIS and its Integration

ACIS Version 1.2 from Spacial Technology (Boulder, USA) is used as the geometric kernel of the product model. It is a modern and quick object oriented geometric modeler which is able to handle very complex shapes. The modeler consist of a boundary representation model which is able to describe non-manifold topology and a wide range of geometric entities like sculptured curves and surfaces (B-splines). The architecture of the ACIS modelling kernel (figure 41) consist of a set of classes to describe topology entities and geometry entities. The application developer can use this modelling kernel like a library and can call either the so-called application procedural interface (api) or directly the methods of the single classes. The class methods are mainly used to retrieve values from this classes (like the root point of a PLANE). Modifications of instances are normally done by api-calls which assure data model integrity and allow an easy way to undo changes within the model. The api-functions include all sort of boolean operations, sweep operations, blending and chamfer operations, sheet and wire operations, and a wide range of inquiry functions. It is possible for the application programmer to develop additional api-functions.

The ACIS kernel does not include any interactive input or graphic output facilities. For test purposes there exists an unsupported test environment called ACIS TESTHARNESS. It consists of a command line interpreter and a simple X-Window based graphic output. The command interpreter accepts commands like:

```
BLOCK my_block WIDTH 100 DEPTH 100 HIGH 50
DRAW my_block
```

The draw command opens a graphic window and shows a perspective wire frame image of the generated block. No graphic based interactions like interactive identification of positions are available.

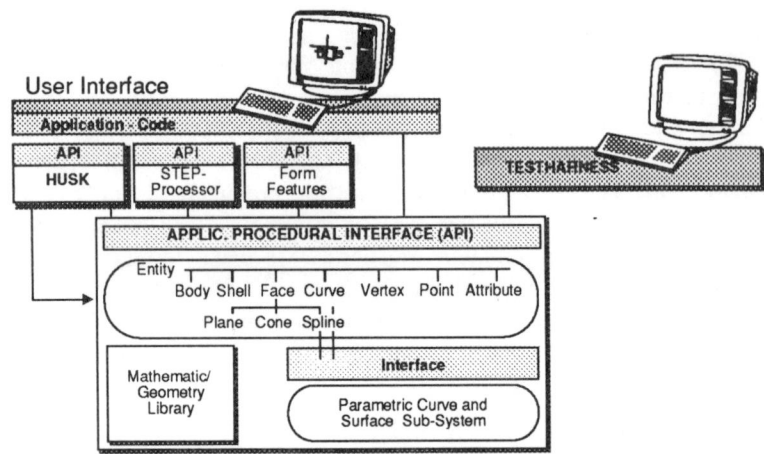

Figure 41: System Architecture of the ACIS Geometric Kernel

For the application programmer exists several ways to integrate the geometric kernel into his application (within ACIS documentation an application is called a HUSK). One way is the specification of additional classes within the geometric kernel. The advantage is that all basic methods like the undomechanism can be used. The disadvantage is that a large set of predefined ACIS methods must be supported (inheritance of virtual methods) even if they make no sense for a certain new class. This type of enhancement is not documented. Another way to integrate the ACIS kernel into an application is the specification of new classes outside of the kernel and the use of ACIS classes as attributes of the application classes.

Within the CACID prototype the second possibility was chosen. This allows us to specify our own basic methods and to create application programs which are mostly independent from future changes of the geometric modelling kernel.

Realised Product Model and Design Space Functionality
The realised product model of the CACID system consists of shape descriptions, form features and design spaces. Form features are objects like a chamfer, a through hole, a plate, etc. The advantage of form features is the possibility to remove, insert, modify shape objects and to define parameters on them. The prototype provides a set of predefined parametric form features and the possibility to define new form features by combining predefined features or the explicit description of a feature by a given shape.

Design spaces are in the first case abstract objects, which are related to a technical solution. They collect a set of constraint descriptions which must be fulfilled within the product model. At the moment these descriptions are textual and shape constraints. A shape constraint is a three-dimensional area (or several areas) where modifications in the product model can take place. Shape constraints can be modelled like other shape objects. They are part of the product model.

If a designer likes to work on a design task, he has to create a new product model version and he automatically reserves the corresponding design space (figure 42). During the design transaction he can modify the complete product model which he has reserved. A 'design space test-button' enables the designer to start a test procedure which test if all the modifications fulfil the design space constraints. A design transaction can only be finished if all design space constraints are fulfilled. The constraint testing in this case consists of the testing of each form feature created during the design transaction against the shape constraints. This is done by boolean operations.

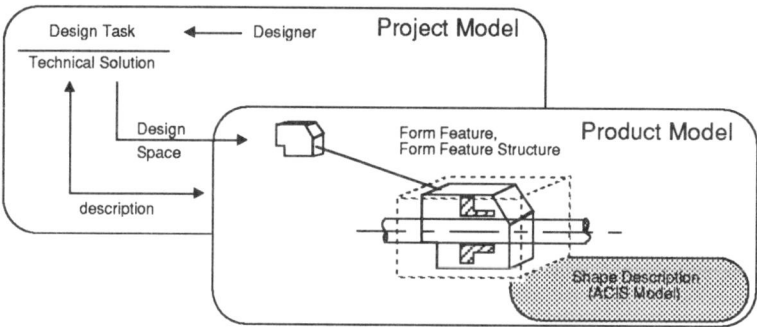

Figure 42: Integration of the Project and Product Model

If several designers are working in parallel, they work in different design spaces. At reservation time point no design space is allowed to 'overlap' another design space. So the merge procedure can be sure that no objects outside of the reserved design spaces are modified during the transaction. This makes the merge process very easy.

Integration of the Object-Oriented Database System
As mention at the beginning of this chapter all changes of project or product model data are directly stored within the common database system of the CACID prototype. This is a feature of the new database technology. This technology does not distinguish between main memory and persistent memory. There exist only one common model (figure 43/1) which is shared by the database and the application. The big advantage is that no longer any load, store or mapping procedures must be programmed within the application. This was a

big advantage within the software development of the first CACID prototype. But a side effect was that the ACIS kernel as part of the product model has to be modified on some places. The reason for this is that all object-oriented database systems demands changes in the inheritance structure or in the allocation procedures of the application.

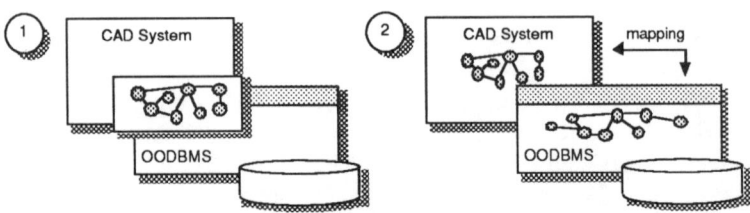

Figure 43: Integration with an Object-Oriented Database System

The need to change a commercial software package was a mayor weakness of the developed prototype for several reasons: Source code and knowledge about the internal structure of the code is needed and complex immigration path to new ACIS versions.

To avoid these disadvantages in the future, the system should use two different models. One for the application and one for the database (figure 43/2). The disadvantage of this architecture is that the mapping algorithm between the two models has to be worked out by the application developer.

5.2.2 Distributed and Open System Architecture

The planned functionality of the system determines what is to be implemented. This is compounded by the decisions of how it is to be implemented. This is defined by the architecture.

The architecture determines which properties the system has in terms of how they are realised. The purpose of the system architecture is to realise the desired functionality in a certain way. The CACID project has set a number of goals which the architecture has to realise.

The architecture decomposes the complete system into several subsystems and describes the interfaces between them. A thoroughly designed architecture helps very much in managing software development and to increase the life cycle of the product.

Goals of the Architecture
The following is a list of the objectives the CACID architecture tries to achieve through it's design. These goals come from the analysis of market needs, and from the chosen functionality of the system as a whole.

Openness

A major demand of customers of CAD software is openness. Openness has several dimensions. It's most important aspect is the ability to integrate with other systems, using their data, or accessing their specific functionality. This is largely determined by supplying the necessary interfaces and by using standards for them. This allows integration on a data representation and interchange level. A further level of openness is the ability to have system's functionality be extended by third parties through the use of functional interfaces. As of these interfaces are still mostly internal and undocumented. For an open system such interfaces must be properly specified and implemented.

For the CACID system functional openness has been set as a target, which means, that all modelling functionality must be accessible. Open data exchange is possible if an accepted standard is able to match the advanced features of the system.

Scaleability

Scaleability stems from the user demand of starting with a simple and cheaper version of the system and continually upgrade it as experience with it grows. It also means that the system can be configured in several ways, so that it can be adapted to different users needs. Scaleability also concerns the hardware. The customer must be able to choose from a range of differently priced hardware, also with the target of low entry level costs and upgrades as demands increase.

The goal of the CACID system is to support different hardware through the use of hardware independent software components and operating system interfaces.

Performance

Architectural decisions influence the performance of the overall system more, than detailed and optimised code. Care must be taken to ensure that system architecture does not prevent the achievement of performance goals.

It is necessary to distinguish between to types of performance: There is the physical performance, in respect to time and space and there is the functional performance of the system.

For the purpose of time performance targets we shall distinguish four levels of response times. These are determined by the frequency of the designer executing the underlying function. The more often a function is executed the faster it must be:

1. Daily operations

   These are functions like start-up of the system or back-up to tape. Response times in the range of minutes are accepted by most users.
2. Hourly operations:

   These are the loading or storing of a sub model or the checking of the work against a Design Space. Response times of up to ten seconds are acceptable.

3. Immediate operations:

   These include the start-up of a modelling function, regenerating a specific display, saving of checkpoint data etc. . Response time must be well below one second.

4. Real-time operations:

   These are operations that are executed with direct feedback, to give the impression of direct manipulation. They include rubber banding, moving and resizing objects and scrolling. Response time must be below 1/25 of a second.

For the purpose of space comparisons we shall distinguish by the number of objects that the system be able to hold in memory simultaneously:

1. Single objects

   These exist usually only once per workstation. They include programs, product models, the display etc.

2. Multiple objects

   These are objects of a complexity, that require the full attentions of the user for their manipulation. Therefore the system only needs to hold a few of them in memory at one time. Examples are completely modelled piece parts, top level display windows, documents etc.

3. Frequent objects

   They are the elementary objects which are operated on, like characters, paragraphs, lines, points, buttons, patterns. They can occur very often. Documents of several million characters or drawing consisting of hundreds of thousands of geometric entities must be supported.

All objects together must fit into the virtual memory of the workstations on the network. Since all objects share the same available space, one group can be traded against the other. However all that is needed, is a upper bound for each group. It is assumed, that no group may consume more memory than the swap space of current workstations. Current hardware limitations suggest this to be one giga byte. Figure 44 gives the expected performance levels of the important operations of the CACID system.

Extendibility

Extendibility defines the possibility to supply more and more elementary functions as the system evolves. Openness in contrast is the ability to add high level functional modules, whereas extendibility makes it possible to add new functions to any module at any level. Extendibility can be achieved if the functions of a module can be designed in an independent way. Then new functions can be added without interference.

Principles of the Architecture

The above goals shall be used through the application of the following principles. They are the basic means to achieve the desired properties of the system and follow the objectives of the design methodology below.

- Allow independent and concurrent development at lower levels of abstraction, after initial analysis is done.
- Allow quick development of a prototype and iterating development steps.
- Allow commercially available libraries to be used.
- Have the support of commercially available CASE tools.

Object Orientation

Object orientation is chosen as the central paradigm of the software architecture. It allows modularisation of the system right down to a detailed level, and is able to model the problem domain very well. Objects occur at all levels of the system. The system's software components are described in terms of objects as well as the data.

Function	Level	Reachability
Start-up	1	+
Loading of model	2	o
Complete redisplay of modified model	3	-
Modification of form features	3	-
Direct manipulation of 3D shapes or form features	4	-
Display of annotations of a design task	3	o
Recalculation of cost model	2	+
Switching from one design space to another	2	+
Move principle element	4	+
Move a node in the Task Decomposition Ed	4	+
Link task and solutions	4	+

Figure 44: Performance Targets of CACID System Functions

Distributed Process Model

The functions of the system are implemented in several separate processes which run in parallel. This allows independent development of each unit and automatically enforces integrity of the interfaces and a high independence. The interfaces between the processes are described through remote object methods.

Model View Controller Paradigm

For each object from the application domain being modelled, three aspects are kept separate.

The data model as such which has no reference to the way it is displayed, and which can only be transformed from one consistent states to another. Intermediate internal states are not made visible, and a transformation may only be applied when all information needed is available. Operations on the model are atomic.

The view of the model, which is not part of the application domain. It is not controlled through the information in the model, but rather through an independent internal state. Transformations of the view, i.e. the way it is displayed, do not modify the object.

The control component. It determines how the user can transform the object and knows about the internal state of a user transaction. It has no information of the contents of the object as a whole, or how it is displayed, but only knows the functions which initiate automatic transformations.

Software Components
The architecture has to take use of high functionality software components into account. These components usually fulfil basic services, like geometric modelling or data access. In this respect a bottom up design is required, to achieve the maximum benefit of the software components. It is not attempted to develop meta-interfaces, which are independent of the used components. Such meta interfaces allow the exchange, of the tools used, but also reduce the functionality that can be accessed. Such components encompass, the geometric modeler, the user interface tool kit, the persistent object database, reusable object container modules and the user interaction state machine.

Elements of the Architecture
The functionality of the system is grouped into different types of software components, which realise common services for each group of functions combined in a module. In general the CACID system is decomposed (figure 45) vertically according to services, and horizontally according to function. Vertical services include: user interface, user interaction model, interprocess communication, editing, modelling and data storage. Horizontal function groups include: project management, task decomposition, geometric modelling, standard parts and technical object editing.

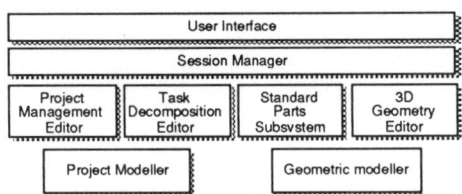

Figure 45: Block Diagram of the CACID Architecture

Editors
An editor incorporates the control and view of a subset of the application domain. In this way there exist a number of editors, each maintaining one type of document representing the technical objects. There is an editor for engineering requirements, for mathematically formulated constraints, for 2D drawings, 3D models, NC generation, task decomposition, version control etc. Each Editor uses the services of the user interface tool kit to implement it's own user interface.

The editor uses the services of a modeler to store, transform and update the applications data. The services of the modeler can be accessed through distributed communication facilities, which allow editors to be implemented as separate processes which can run anywhere in the network or can be scheduled to different processors.

Modelers

Modelers supply basic computations on the mathematical data model. The CACID system currently uses mainly one modeler, which is the geometric modeler ACIS from Spatial Technology Inc. This modeler maintains the boundary representation structure of the bodies being modelled with their topology, and supplies functions for their creation, transformation and modification. On the top of this modeler a form feature model has been implemented, which maintains the parametric structure of the geometry in a CSG tree.

Communication

Each component of the CACID system that is modelled as an independent process, exchanges information via a communication facility. This facility is implemented as the NesCAD Interface Protocol. This protocol enables CACID sub processes to send arbitrary objects, data streams or commands to other processes.

User Interaction

The CACID system follows an operator oriented user interaction model, where interactions are described by finite state automaton. Building of user interactions, and execution of the automaton is supported with the 'Algorithmic State Machine' ASM.

User Interface

User interaction in CACID relies mainly on the direct manipulation of objects. In order to give the user as many options as possible to act on the object a tool kit is used which easily allows the building of specialised widgets. This is facilitated through the use of the InterViews tool kit supplied from Stanford University. It is used to build the pictogram menus and text editing fields in the system. The specific and extended functionality is available from the UMI library developed in CACID project.

Persistent Data Storage

Persistent data is all data, that lives longer, than one user session. Such data is stored in the product, or project model. It is desirable to store all objects one to one onto the disk. To achieve a very powerful tool in the form of an object oriented database is used. The product chosen for this purpose is the Objectstore system from Object Design Inc.

Display

The display component of the system is used to view the geometric models. For the display of the 3D models of the NesCAD 3D system the tool kit 'Hoops' from Ithaca is used. It is able to do all shading zooming, panning and rotating of the model independently.

Distributed System

In order to extend the functionality of the geometric editing component, a commercially available CAD system was integrated into the prototype. Using this CAD system a distributed and open architecture was realised to combine the components. The functionality required to model the concurrent and integral design process was grouped according figure 46 and 47.

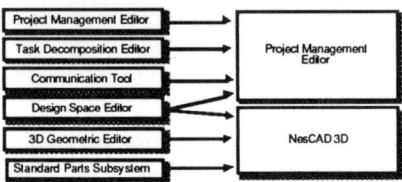

Figure 46: Functional Components of the Distributed System

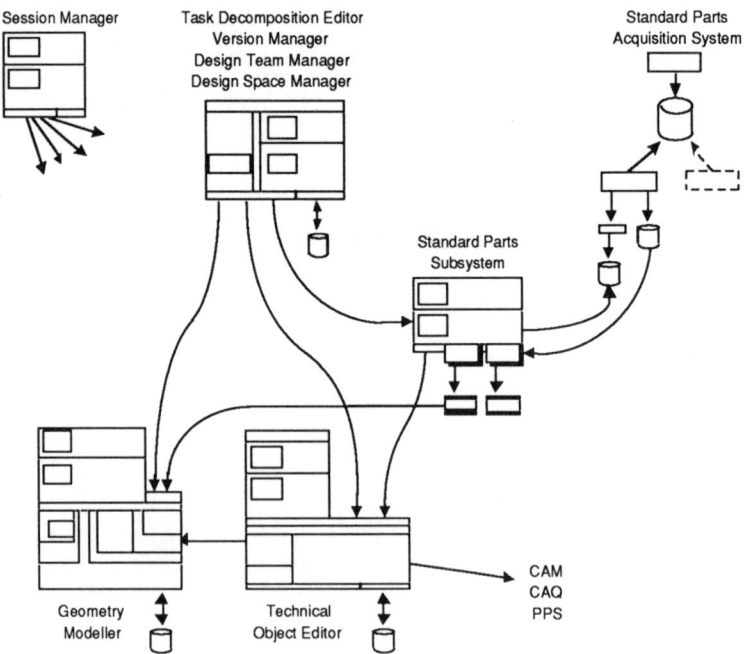

Figure 47: Architecture of the Distributed System

5.2.3   The Standard Parts Sub-System

This chapter gives an overview of the standard parts sub-system and its integration into the CACID system. The standard parts sub-system is composed of:

- the standard parts sub-system itself, which runs during the design process in order to allow the CACID's user to retrieve one standard parts from the library, to insert this part into the geometry database and into the technical objects library,
  Thus from the geometry database, the part can be displayed on screen, and from the technical objects library, technical evaluation can be performed using design criteria.
- The data acquisition system, used by standard data suppliers. This system can append new data to the standard part library, or update existing data.

The standard parts sub-system for CACID has been studied and developed according to the following criteria:

- fulfil the specific requirements of CACID, for concurrent integral design,
- be compatible with existing data libraries,
- follow as close as possible the standardisation works undertaken in the field of the parts library at CEN and ISO (CADLIB),
- consistency with CADLIB,
- compatibility with existing library format.

The CACID project was running while standardisation works were performed both at the European level and at the international level. The following document, studied at the CEN and ISO, was considered:

- CAD/LIB Logical description format of supplier library.
  (CAD/LIB Part: 24 - E, CEN/CLC/IT/CAD-LIB/WG4 N11.)

According to this document, a STEP file should be used (with new functionalities) to describe and exchange a library, using the EXPRESS language. Then this library is used by a library management system to select parts from the library and to produce views of these parts into the CAD system. There exists, at the moment, no tools supporting the exchange of libraries defined in this way. It was not in the scope of the project to develop an EXPRESS compiler, and a full consistency could not be achieved with CADLIB.

- the CADLIB functionalities (model descriptions, rules and constraints on the model, dictionaries, external representations, geometry programs) have been implemented either with C++ methods or with relational database tables (selection programs),
- the CADLIB object oriented structure for parts libraries has been implemented as C++ declarations of hierarchy of classes,
- a standard interface for graphical representation has been used.

The main problem arises with the DIN's libraries as described in DIN 4000 and 4001 standards and the DIN technical report N14. These libraries are not fully compatible with CADLIB concepts, but it has appeared that most part of the data could be used within a CADLIB approach: the characteristic data, providing the file format compatibility, and the geometry programs providing the graphical interface compatibility. Therefore the implementation choice were the following:

- DIN format for the characteristic data (attributes),
- VDAPS-3D for the graphic interface.

The standard parts sub-system (figure 48) should allow the user to select a part family. This can be done with the hierarchical structure of part families, or with design criteria. The user should be able to define an occurrence of a part family, that is to give values to the parameters of the part family ($D = 10$, $L = 100$, ...). Constraints should help and/or control the user input. It should be possible to produce different views of the occurrences of a part family. Three types of views are considered in CACID: principle element, rough shape and detailed model.

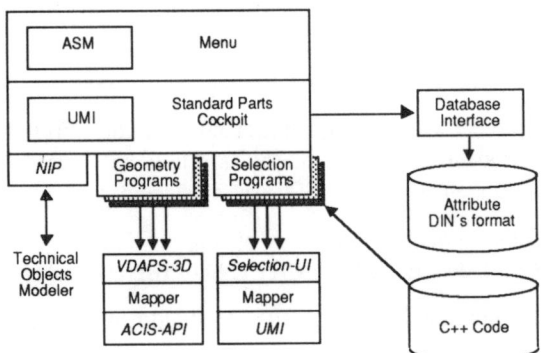

Figure 48: General Architecture

The data acquisition tool is composed of a supplier database, a data acquisition tool, to input data in the data base (figure 49), and a procedure to extract and assemble automatically data to produce a library for CACID.

The schema of the supplier database has been studied to be consistent with the standardisation work in progress at CEN and ISO. This database schema has been implemented under the INGRES environment: relational database management system, high level language.

The acquisition tool, which allows to input in the supplier database the object oriented description of a part family, has been implemented under INGRES. Procedures to extract data from the database and to assemble data to produce automatically a library for CACID, written in C++, has been developed with INGRES.

Using the formalism and the methodology of CACID the conceptual data model is defined on several levels: General integration of the standard parts sub-system as part of technical object and product modelling, and modelling of the standard parts sub-system, using the same concepts of modelling as for the other product model entities, and especially the same parameter modelling.

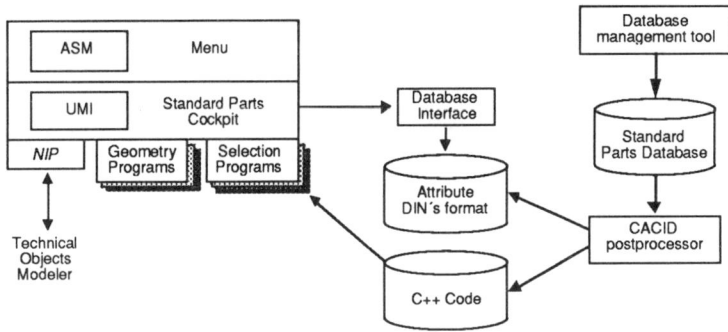

Figure 49: Data Acquisition Tool

The integration of standard parts into the CACID data model is achieved by the definition of the class STANDARD_PART. It is a sub-type of the class PRODUCT_MODEL_ENTITY (figure 50)

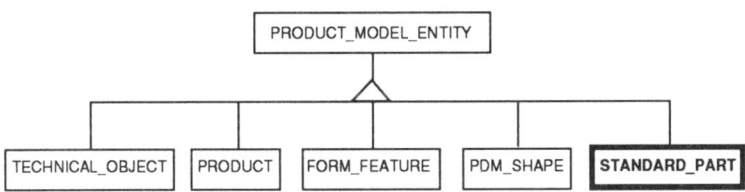

Figure 50: Standard Parts Sub-System Integration

The standard part data has been modelled from the models developed so far for products and technical objects. The first schema used for the standard parts modelling is the general concept of class and class occurrence modelling, which applies to the TECHNICAL_OBJECT class, or to the FORM_ FEATURE class, for instance.

The second schema used for the standard part modelling is the PARAMETRIC_ENTITY model. Figure 51 shows the standard part data model. Due to the use of general schema of CACID, the addition of this model to the TECHNICAL_OBJECT will allow to use the already developed tools for the new model. As a matter of fact, this addition of a new model is a functionality available to the existing system.

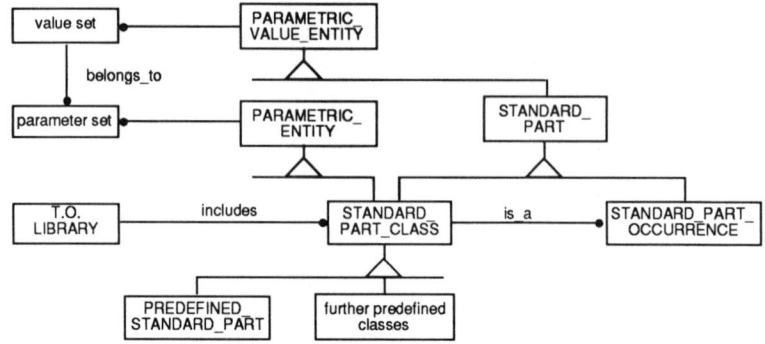

Figure 51: Standard Parts Model, using General Structure of Product Model

The standard part sub-system is composed of a library, which describes standard parts as classes derived from the PRODUCT_MODEL_ENTITY_-CLASS and from the PARAMETRIC_ENTITY class. All the methods available for the technical object library apply to the standard parts library.

As for the other PRODUCT_MODEL_ENTITY, an occurrence of a standard part will be managed by the specific methods of the STANDARD_PART classes. These methods will provide the way:

- to give values for an occurrence of a class to the parameter set of the corresponding class, according to the rules and constraints which apply to the class. The dialogue with the user is set through the UMI of CACID,
- to realise the graphical view of the occurrence within the graphic model of CACID.

The standard parts sub-system, derived from the conceptual models of CACID is composed of C++ classes which described the part families, and a library interface which can create an instance of a class and can call methods to produce views into the CACID system.

The C++ classes encompasses the data structure using the object oriented capabilities of C++. Methods are defined in the class structure to provide selection programs, for the selection of a part family and for the selection of an occurrence of a part family. Design criteria can be used for the selection of a part family. View programs are provided for the production of views into the CACID system. For the geometry view, these methods call FORTRAN routines which call the VDAPS 3D geometrical interface. Other views can be used for technical evaluations. The selection programs access tabular data to help the user with the choice of standard values for part family parameters. Tabular data is in DIN format, and is accessed through a database interface.

The part families are described as a hierarchical structure. A part family inherits all methods and data from the class STANDARD_PART_CLASS (figure 52).

Three levels are distinguished (figure 53):

- the top level, which is the entry point in the hierarchy,
- the intermediate levels, for classification purpose,
- the basic level, which can be associated with a real part.

```
/* The class STANDARD_PART_CLASS is the top of the hierarchy of the
standard parts subsystem.
Each class which describes a part family is derived from this class.
*/

class STANDARD_PART_CLASS {

/* CACID virtual method */
virtual char* class_name();

/* list of parameters (to be implemented in the subclasses)   */
/* selection programs                                         */
virtual get_value_pe();
virtual get_value_rs();
virtual get_value_dm();

/* geometry programs                                          */
virtual realize_view_dm();
virtual realize_view_rs();
virtual realize_view_pe();

/* design criteria methods                                    */
virtual design_criteria_pe();
virtual design_criteria_rs();
virtual design_criteria_dm();
};

/* list of parameters should appear in the following way:

type        attribute;
CHAR        attribuite_name;
SPVALUE     attribute_spavlu;

where
type stands for DOUBLE, INT or CHAR*.
attribute is the internal name of the parameter (A1, A2, ...)
attribute_name contains the external name of the parameter,
and is initialized by the constructor.
attribute_spvalu is used with the interface.
*/
```

Figure 52: STANDARD_PART_CLASS Description

At the top level, one can get views on the part family as a 'principle element' representation. At the intermediate level, one can get views on the part family as a 'rough shape' representation. Principle element representation is inherited. At the basic level, one can get views on the part family as a 'detailed model' representation. Principle element and rough shape representation are inherited.

The library interface functions are:

- PART_CHOICE: choice of a part family
- VIEW_CHOICE: choice of a view and
- GET_VALUE: get value for the parameters of an occurrence
- REALISE_VIEW: realise the view into CACID
- NEW: creation of an occurrence of a part family

The GET_VALUE method assigns values to an occurrence p_spc of the abstract class STANDARD_PART_CLASS. This method calls the get_value method of the part family class and the get_value methods of the upper level part family classes which are inherited. The REALISE_VIEW method realises the geometrical view of an occurrence p_spc of the abstract class STANDARD_PART_CLASS. This method calls the realise_view method of the part family class or of the upper level part family classes which are inherited.

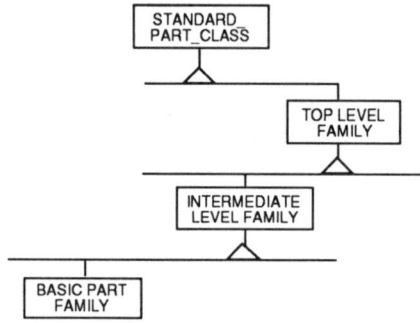

Figure 53: Hierarchical Structure of the Part Families

The NEW function creates an instance of class which describes a standard parts family. Such a class is derived from the abstract class STANDARD PART CLASS. Input parameters are the domain_code and the model_code. For each domain, there exist a function whose name is NEW + library_code and which performs the creation. The general NEW function calls the specific NEW function according to the value of domain_code.

The use of design criteria have been studied with the end user's of CACID (Dessindus) during a preliminary design phase, and then implemented in the standard parts sub-system.

A preliminary design has been performed with Dessindus. The objective of this preliminary design was to study and define how Dessindus uses design criteria during the design process. Some catalogues were studied. These studies and the knowledge of Dessindus in the field resulted in the following results:

- some criteria being defined, the evaluation of this criteria can be performed either by a method or by a person.
- a criteria can be used for the following processes:
  - choice of a part family,
  - choice of an occurrence of a part family by a method,

- evaluation of an occurrence of a part family by a method.
- no evaluation method could been defined for the CACID project.

• the use of design criteria for the selection of a part family was considered as most important for the designer by Dessindus.

Following the results of the preliminary design phase, the use of design criteria for the choice of a part family, using a person for the evaluation of the design criteria was implemented in the data acquisition tool and in the library interface of the standard parts sub-system cockpit.

The possibility to define a design method for an occurrence of a part family was implemented in the data a acquisition tool. But as such method has not been implemented in the CACID system, this possibility has not been implemented in the library.

An overview on the implementation of the library and the links with CACID interfaces is shown in figure 54.

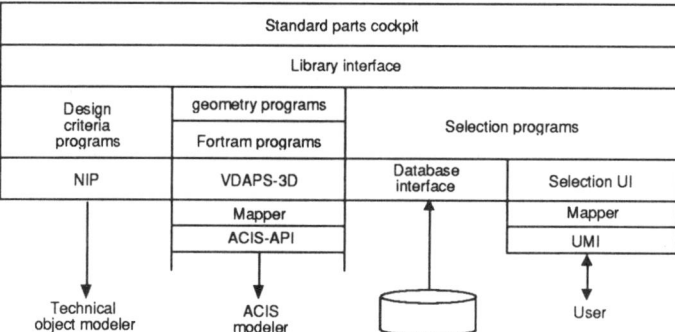

Figure 54: Implementation of the Library and the Links with the CACID Interface

Figure 55 presents parts the standard part sub-system user interface, which was developed as a prototype. It consists of a main icon menu, where the user can select a part family. The other windows are used to select values for the part attributes.

The Data Acquisition Tool

The data acquisition tool is intended to be used by standardisation bodies and catalogue suppliers. The tool implements all the concepts defined within the CADLIB standardisation works. The data acquisition tool is composed of:

• a database,
• a data management tool, to append data or to update the database,
• a C++ post processor which retrieves data from the database and builds automatically the C++ library. This C++ library will be compiled and link with the other elements of the standard parts cockpit.

A CADLIB post processor is also available, which produces files using the EXPRESS language.

The evaluation of criteria for an occurrence of a part family is done following the general conceptual schema of CADLIB. Evaluation can be considered as a functional view of a part family, and it exists a functional model of this part family for this view. This is the same as for the geometrical representation. The first data acquisition tool was hard coded for the geometrical view of part family. This first version has been rewritten to handle several functional views.

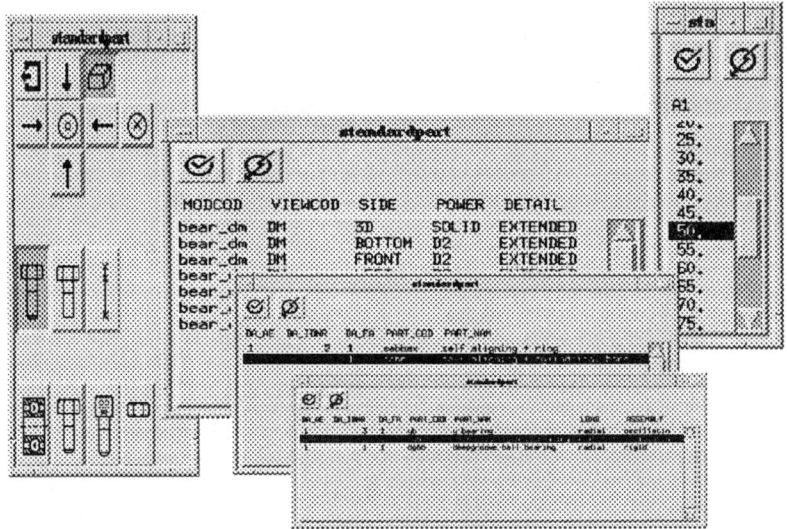

Figure 55: Standard Parts System User Interface

The first version of the standard parts sub-system provided a hierarchical choice for part family. The final version implements the use of design criteria for the selection of a part family. This is implemented within the PART_-CHOICE ( ) method of the library interface.

One high level part family being known, this method displays a list of low level part families together with the associated design criteria values. The end user reads these values and choose a part family.

6 Evaluation of the Prototype

In the evaluation phase of the CACID project, the user's task is to test the supplied CACID prototype by using the tools which make up the prototype to carry out new projects. This mainly concerns the use of the task decomposition editor and the design spaces which are managed by the system. As in the standard part library just a small number of components are available, this tool was not tested in industrial use.

6.1 Example Used for Drawing up the Test Report

The example selected to test the prototype is the design of a new range of unwinders. This example is particularly interesting because of the design method chosen. In order to avoid any exchange of ideas, the prototype will be developed at two different sites.

After the specification have been received, two teams which have absolutely no contact with each other will develop the same assembly. In this way, the differences arising from the use of CACID tools will be brought out more easily.

6.1.1 Description of the Design Task

The work to be performed concerns the design of an independent machine element used to unwind paper reels (figure 56). The design must take into account the development of a winding function within the element. There must be two run off speeds, one of about 50 m/min and another of about 100 m/min.

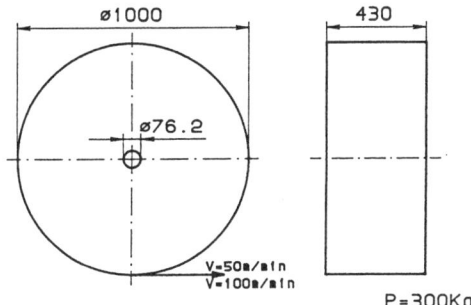

Figure 56: Paper Roll to Unwind

As the equipment could be used in computer centers, noise levels and aesthetics must also be taken into account. The aim is to design an element which could be sold for less than 100.000 FF.

6.1.2 Research of Alternative Solutions

Here is how the research of a problem solving principle is carried out with the actual method.

After the constraints imposed by the specifications had been analysed, a sketch which gave a general idea of the element to be designed was done (figure 57). The sketch was then distributed to the members of a group which had been formed earlier. After one day of research, the sketches were put together, along with the principle diagrams and, in some cases, a description (figure 58).

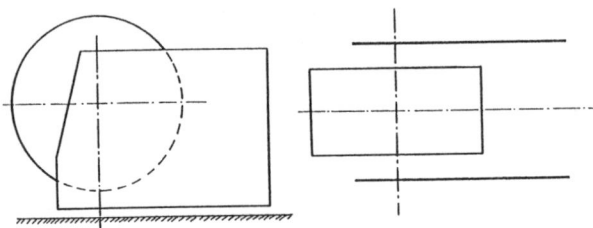

Figure 57: Basic Sketch for Research of Alternative Solutions

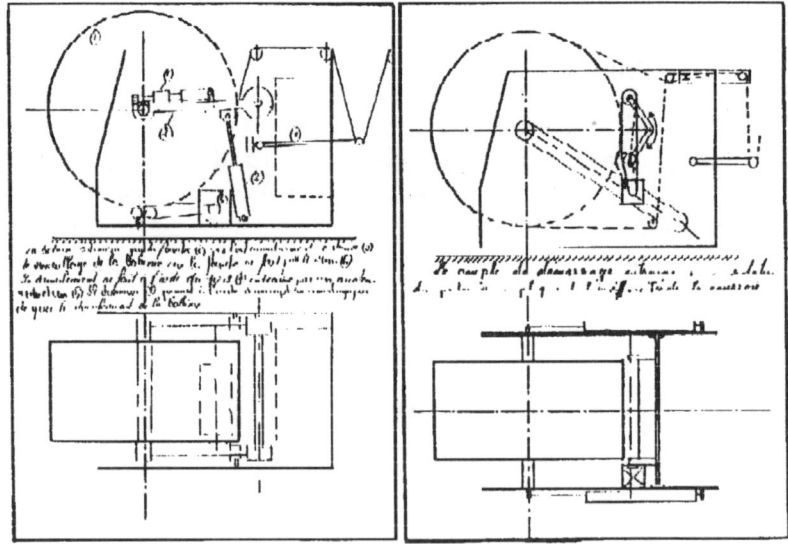

Figure 58: Results of Research of Solutions

Thanks to the number of participants, this method enables us to quickly obtain several solutions. The choice of the solution to be developed was made once all the propositions were discussed (figure 59).

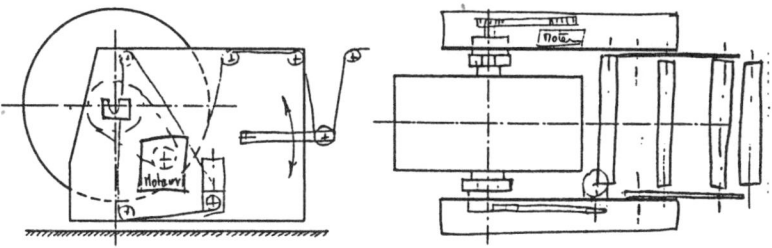

Figure 59: Solution to be Developed

One must note that with this method, several participants can develop similar solutions. As there is no structure to organise and orient the solutions and classify the propositions, we cannot ensure that all the solutions are analysed.

This is different with the CACID tools. During the principle research stage, the task decomposition editor (TDE) was associated with 2D drawing software (figure 60). The research was done by only one draughtsman. The use of the TDE enabled us to have a highly organised research.

As the solutions are classified progressively, one can have an overall view of the research at any time. Constant access to the data related to the solutions means that one can do the research at different times.

In this case, it is much easier for another person to carry on the research. As the simultaneous display of several principles is impossible, the draughtsman has to store them in order to make a choice. The research took about twenty hours, divided into two-hour sequences.

The use of TDE mainly avoided developing the same solution twice. The greatest constraint on the draughtsman was the fact that he had to recapture data in the TDE windows.

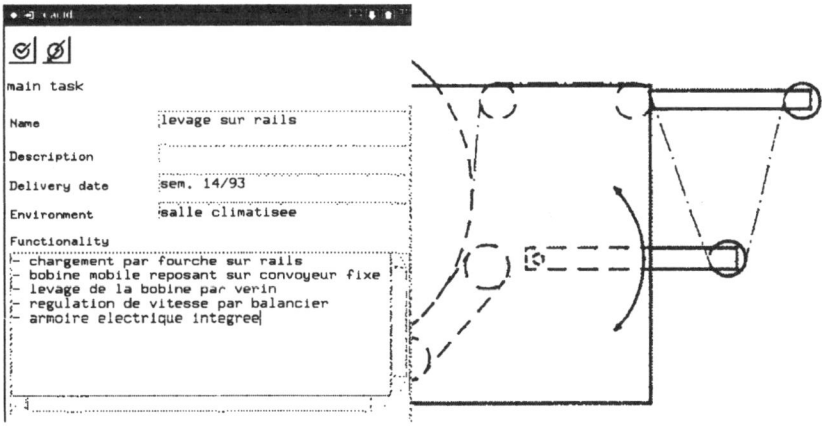

Figure 60: Results of Research with Help of CACID Tools

As the two groups came to similar technical solutions, we selected one of them, and developed it. The same proposition was given to the two work groups so as to facilitate comparison. This early stage of the design process brought out the usefulness of a tool which could organise and supervise the progress of the research at any time. The highly organised presentation of the different solutions helped us to save time by using the TDE. We could save even more time if sketches could be joined to the solutions.

As the work done is stored, it can be taken up by another person. With the conventional method, the absence of a draughtsman working alone could have serious consequences. Even if the TDE does not greatly increase immediate productivity, it does assure more continuity in the work.

6.2 Description of the Work Methods

Once the functions of the unwinder are defined, they are currently used to distribute the work among the draughtsmen (figure 61) and each designs the sub-assembly he has been allocated. Element interconnection can only be secured trough a high level of communication between the participants. A substantial change in a sub-assembly often requires changes in the plans of several draughtsmen. One of them is asked to plot the overall plan in order to ensure the assembly of the sub-assemblies. This plan is updated as the work progresses. Once all the designs have been done, the group chief marks and lists the parts.

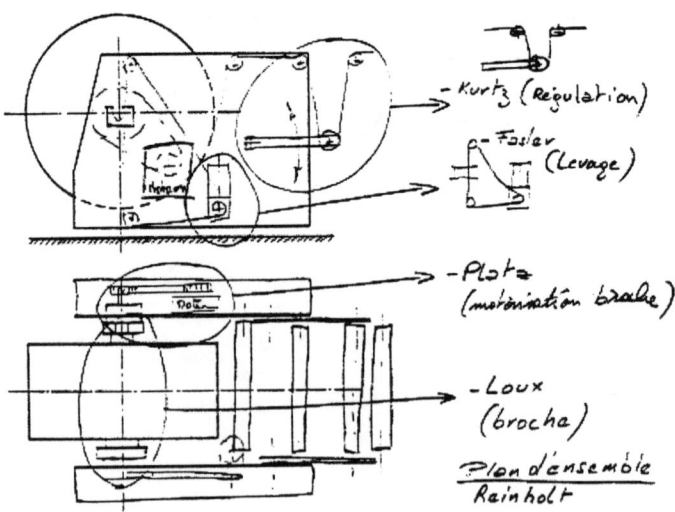

Figure 61: Distribution of the Work among the Draughtsmen

The execution plan plots are done at this stage of the project. The draughts-men take care of the detail plans related to the sub-assemblies defined by them. In this way, the chances of errors arising out of part connections are reduced.

Using the CACID tools, this phase of the design is organised in a different way. With the help of the know functions of the unwinder, the group chief created design spaces which are allocated to the draughtsmen (figure 62). A group of three draughtsmen is formed to carry out the project. As the overall view is done automatically, each of them designed one sub-assembly. The work progresses quite steadily with design spaces in which the function to be developed does not interfere much with the others. If a design space contains an element of the machine which is often used by all members of the group - e.g. a frame plate - successive merge operations considerably slacken the speed of the work.

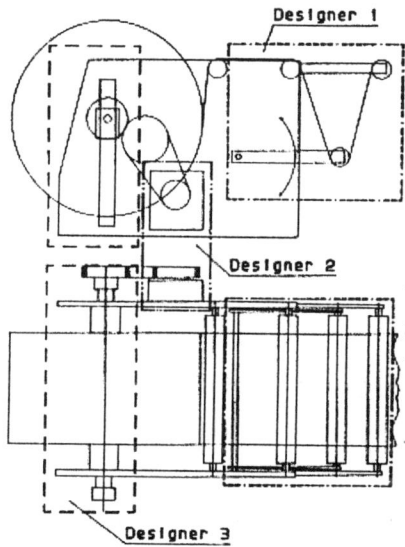

Figure 62: Attribution of Design Spaces to Designers

We did think of putting the whole unwinder in just one Design Space, but as the other draughtsmen could not have access to their design space, there was just one person on the job, and the progress was slow. As marking could not be done with the 3D software, it had to be done with the 2D software. If the list of parts had been generated directly, we could have saved time by linking the parts of the unwinder with the TDE information tables. The transfer of 3D elements to 2D helped us save 30 % of the time taken for this stage of the work. This method was more efficient than recovering elements from 2D plans.

The problems posed by the current method often arise from the need to draw some elements again when they are altered. When it comes to avoiding unnecessary operations, the CACID prototype fulfils its functions completely. Time losses with CACID are often related to slow hardware reactions. This slowness also means that the frequent design space transfers are tedious.

As the system does not detect the interference between the parts, it is up to the draughtsman to carry out this task, as is the case with the actual method. Even if the list of parts is not automatically drawn up, the 3D list of parts helps avoid omissions. Time is saved most while making the execution plans. In this phase, speed and reliability have been improved.

6.3 Problems Faced in Working with the Prototype

While using the TDE, we came across the following problems :

- many of the signs concerning the cursor do not match the messages they are supposed to be sending, e.g. a waiting period ought to be represented by a clock or an hourglass, but the cursor does not change
- one no longer has an overall view of the pyramid while doing the design of a standard sub-assembly. Due to TDE's need for screen space, one cannot see the plan under way (figure 63)

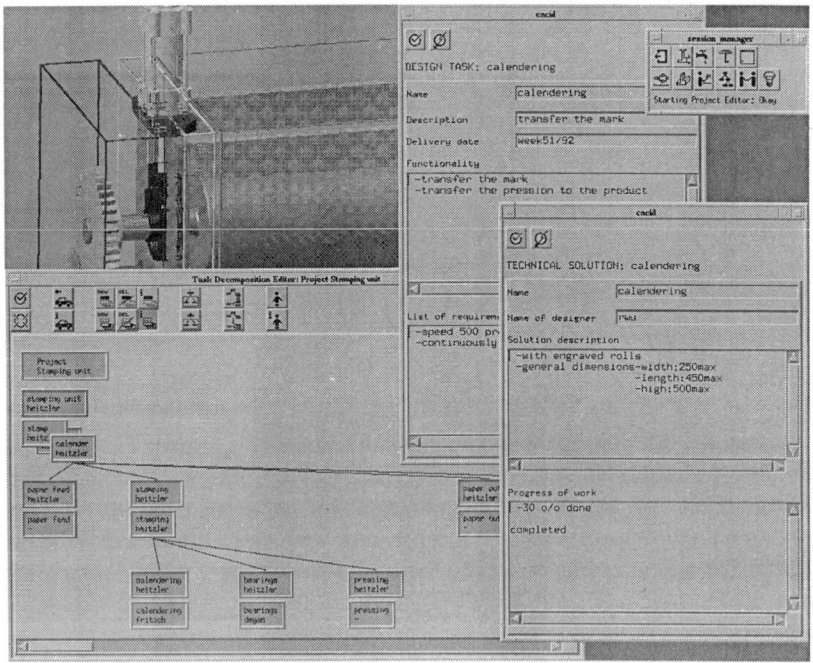

Figure 63: Example of Screen crowding by TDE

As TDE has been developed in order to assist draughtsmen, it ought to be suited to their needs. It is not possible to use TDE rationally, as the technical solution and design task information windows cannot be personalised. The data concerning a technical solution idea is not the same as that concerning the description of a basic component (e.g. bearings, mechanical parts which are to be machined, etc.) and ought to be able to display this difference (figure 64).

If one could have specific tables for the basic components, the system could automatically generate the list of parts (figure 65), and the draughtsmen would not have to capture all data twice.

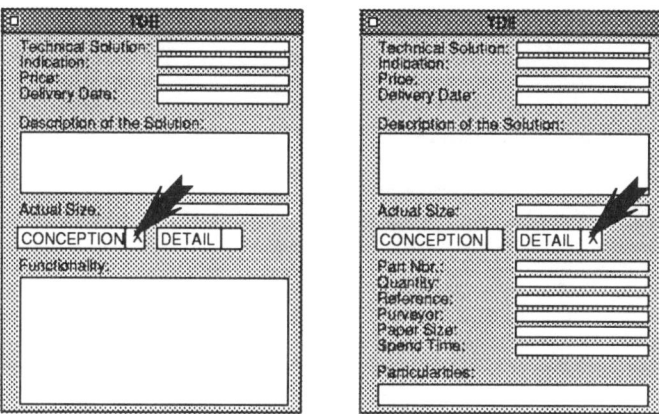

Figure 64: Proposal for TDE Display Windows

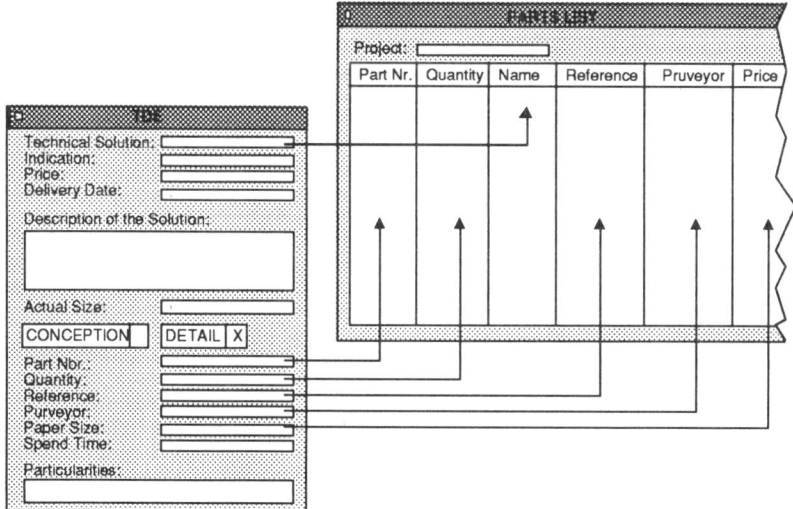

Figure 65: Relations from TDE to Parts List

In the case of a section which describes the constraints given by the specifications, technical solution data is the same whatever the solution may be, and data recapture is responsible for a great loss of time.

Unfortunately, the method used to display the various functions of CACID does always not take into account the fact that it is used on a screen which is primarily used for drawing.

The need to open several windows, where one window is used to select a function and another to enter the data (figure 66) makes access to the drawing difficult.

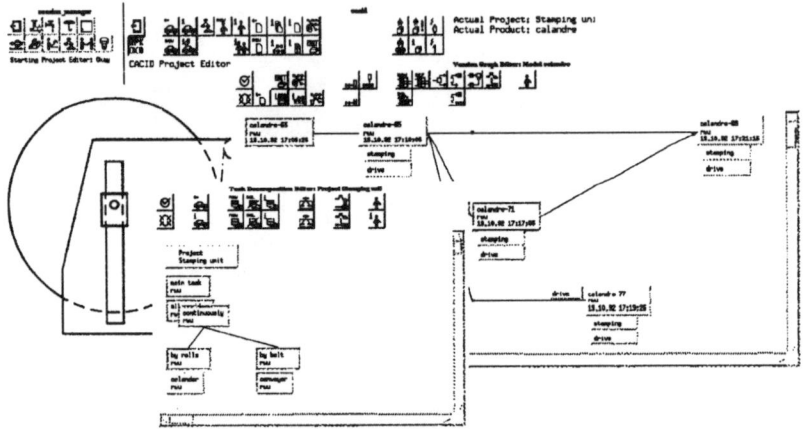

Figure 66: Example of Screen Crowding by Version Graph Editor

As CACID needs a lot of power, which is not always available, it would be wise to tell the draughtsman to wait before sending the next order; when a function is selected too soon, the prototype tends to 'crash'.

As space is not limited for the draughtsman during the designing, it is very unpleasant when one realises that while testing the design space one cannot save the work done. After allocating a more suitable design space, one ought to be able to load a listed assembly in order to replace a standard model.

There should be a message showing the changes in the status of the version graph editor. This message could be textual, or a picture flashing in the version graph editor menu. Such a message would not disturb the draughtsman.

The method which is presently used by CACID to manage the design spaces does not allow us to detect collisions between parts. The only errors shown are those which occur when objects exceed the design spaces.

The accuracy and reliability of the prototype came to us as a pleasant surprise. The only errors were those following user errors. However, it would be useful to be able to force 'merge' even when the design space is exceeded.

6.4 Suggestions for Improvements

The preliminary test period allowed us to highlight the first problems with intensive use.

Search and Classification Problems

During the principle search phase, all the proffered solutions are examined by the draughts person. The technical solutions not adopted are often not profoundly developed, and little written documentation remains to allow their re-use. The principle search task currently makes demands on the draughts person's memory and obliges them to re-invent certain principles. The system can provide considerable help at this level by storing solutions with their descriptions.

The problem that remains to be resolved concerns classification of solutions. While the CACID system is extremely effective from the viewpoint of storage, it is much less effective in providing assistance with design by proposing standard solutions.

The current classification system makes demands on the draughts person to recall a principle already developed for another project in order to re-use it. It should therefore be provided with a library for storing mechanical engineering solutions that the system could suggest. Classification should be keyed on the technical function fulfilled by the principle stored.

Generally-speaking, the projects and principles classification system lacks sophistication. While it should be possible to administer principles within the library, projects require a more flexible classification system. The current classification system does not allow management of projects, and is not designed for setting-up a hierarchy of them. The tittle of a project can only be modified by a programmer. It should be possible to set-up a search for projects according to constraints stipulated in the specifications (Example paper machine: the system should proffer all projects in which the subject was working with paper).

Incorporation of Calculations

At the beginning of the design phase, the draughts person needs to perform calculations to verify the various parts of the mechanism. At the present time, these tasks are carried-out away from the workstation, and necessitate a survey of all dimensional parameters on the plot performed. The ability to do these calculations with the system would improve the precision of results and avoid interrupting the design process.

Using standard principles already stored by the system should provide information concerning the cost of an assembly employing these solutions. This cost can be approximate for the principle search and preliminary study phase, and can be more precise for the portion involving standard parts drawn from the library.

Suggestions for the Improvement of the CACID Prototype

In order to improve the CACID prototype, complementary functions must be developed. Indeed, it would be useful to have a test tool which could increase a design space along with the external contours of a part. This design space should later help us detect interference between parts. The used space function must be developed, thanks to which rough models of mechanisms could be speedily built and defined. In this case, one could compare the function to the conventional manual sketches of mechanisms. The advantage of this approach is that the design space could be easily distributed between the draughtsmen, and each used space would be given a specific function in the overall mechanism.

It would be helpful if the design of the various functions could be reviewed so as to have something more compact, both in terms of screen display and in terms of storage space.

In view of the present condition of the prototype, it seems to us that the need for a second screen during the implementation of CACID is unfortunately inevitable if sustained productivity is required of the draughtsman.

6.5 Conclusion

During the test period, the CACID prototype was used for several projects. From this experience, we can judge the capability of CACID. While smaller projects are easily done, larger designs are more difficult. The problems are mainly related to the hardware used and the 3D software to which the tools have been adapted.

We could not say that we saved time by using CACID. Above all, the use of CACID made us organise our projects better. Often, the real time gains are only seen when there is a problem or when a similar contract is executed. We liked the performance of the system. However, smaller design offices could not put it to commercial use just yet. The main reason for this is that the tool would not be profitable, as the extra cost of investment could only be compensated by a 50 % increase in productivity, and this is far from being true at the present moment. We do think we will continue using the CACID product, but for tasks which are suited to its functions. We hope it will be developed further, as the tool is useful in many ways. Its use could be envisaged by design offices having large financial resources, but the smaller ones will have to wait for a year or two before buying powerful hardware at a lower cost. As the prototype is already reliable, it would only be necessary to add functions which could make it more efficient. Once this is done, we hope this product will be commercially successful.

7   Conclusion

Design methodologies describe the stages of the design process as a step by step process. They do not make any proposal how the work of a designer or a design team should be organised or managed to get good technical solutions in a short period of time.

Within the CACID project concurrent design techniques used by small and medium sizes enterprises are analysed and the results are documented with the Structured Analysis and Design Technique [2,18].

To support concurrent design, information about design task distribution, design task requirements and constraints, the history of the design process and its results, and standard parts are needed. To handle and manipulate all this data within the CACID project a new type of CAD system was developed. The system was completely developed in an object-oriented manner and includes some of the newest CAD software modules like i.e. the ACIS modeler.

The company Nestler ensured that the system architecture is very modular and can be easily extended by further tools. One commercial tool which was adapted and included into the system was the 3D product modeler from the company Strässle.

The evaluation of the developed CACID system showed that engineers can use the system and are very interested in topics like the description of the design task distribution and the distributed management and access of their data. The evaluation also showed that designers have special requirements to the quality and speed of the user interface. The evaluation phase showed that the development of a 100% CAD system which fits all the ergonomic and functional requirements of the user is very hard and needs a lot of man power. In addition to the concurrent design functions the user needs especially calculations routines, drawing generation mechanisms, parts list, etc. To realise all these functions was not possible within the project duration.

The CACID project led to a set of products. Some of them are already available on the market. In general it can be stated that the project realised ideas - like product model versioning, design task distribution, storage of design requirements and constraints - which will be functions of all future CAD systems.

Bibliography

1. Adler, R.E.; Ishii K.: DAISIE: Designer's aid for simultaneous engineering, Comp. Eng. Proc. Int. Comput. Eng. Conf. Exhib. Vol. 1 (of 2), Proceedings 1989

2. Chevalier, A.: Guide du dessinateur industriel, Hachette Technique, Edition of 1989, Chapter 6.2: Analyse fonctionelle descendante

3. Cutkosky, M.R.; Tenenbaum, J.M.: A methodology and computational framework for concurrent, product and process design, Mechanism and Machine Theory, Vol. 25, Iss. 3, 1991, pp. 365-381

4. GARM, General AEC Reference Model, TNO institute for building materials and structures, August 1989

5. Grabowski, H.; Rude, S.: Methoden zur Lösungsfindung in CAD-Systemen, International Conference on Engineering Design, Budapest, 1988

6. Hansen, F.: Konstruktionssystematik, Berlin, VEB-Verlag Technik, 1965

7. Hansen, F.: Konstruktionswissenschaften - Grundlagen und Methoden, München, Wien, Hanser-Verlag, 1974

8. Hubka; Andreasen; Eder; Pighini; Schlesinger; Wyss: Fachbegriffe der wissenschaftlichen Konstruktionslehre in 6 Sprachen, Schriftenreihe WDK 3, Heurista Verlag, 1980

9. Jakiela, M.J.: Concurrent engineering with suggestionmaking CAD systems, Results of initial user tests, Mechanical Engineers (ASME), New York, NY, USA, 1989, pp. 223-230

10. Koller, R.: Eine algorithmisch-physikalisch orientierte Konstruktionsmethodik, VDI-Z 115, 1973, pp. 147-152, 309-317, 843-847, 1078-1085

11. Koller, R.: Konstruktionsmethode für den Maschinen- Geräte und Apparatebau, Berlin, Heidelberg, New York, Springer-Verlag 1976

12. Lu, S. C.-Y.: Computer-based environment for simultaneous product and process design, Mechanical Engineers (ASME), New York, USA, 1988, pp. 35-46

13. Pahl, G.; Beitz, W.: Engineering design, Second Edition, Springer-Verlag, 1984

14. Pennell, J.P.; Winner, R.I.: Concurrent engineering: Practice and prospects, IEE Global Telecommunications Conference & Exhibition, GLOBECOM 1989, Part 1 (of 3), Dallas, TX, USA

15. Penell; Winner; Bertrand; Slusarczuk: Concurrent engineering: An overview for Autotestcon, IEEE Service Center (cat n 89CH 2568-4), Piscataway, NJ, USA, 1989, pp. 88-99

16. Raab, H.W.: Rechnervernetzung und reduzierte Fertigungstiefe steigern die Produktivität, Sonderteil in Hanser-Zeitschriften, März, Carl Hanser Verlag, München 1991, pp. 16-18

17. Rodenacker; Clausen: Methodisches Konstruieren, 2ed, Berlin, Heidelberg, New York, Springer-Verlag, 1976

18. Ross, D.T.: Applications and extensions of SADT, IEEE Computer, April 1985, pp. 25-33.

19. Roth, K.: Gliederung und Rahmen einer neuen Maschinen-Geräte-Konstruktionslehre, Feinwerktechnik 72, 1968, pp. 521-528

20. Roth, K.; Franke, H.J.; Simonek, R.: Algorithmisches Auswahlverfahren zur Konstruktion mit Katalogen, Feinwerktechnik 75, 1971, pp. 337-345

21. Rumbaugh, James: Object-oriented modelling and design, Prentice-Hall, 1991
22. VDI Guideline 2221: Systematic approach to the design of technical systems and products, Düsseldorf, VDI-Verlag, 1987

Publications from the CACID-ESPRIT Project

- Anderl, R.; Malle, M.; Schmidt, M.: Concurrent engineering based on a product data model, proceedings of the conference CALS Europe 1992, 16.-18. september 1992, Paris, Editions Hermès, p. 145-154

- Grabowski, H.; Rude, S.; Schmidt, M.: Entwerfen in Konstruktionsräumen zur Unterstützung der Teamarbeit, in Scheer, A.W., Simultane Produktentwicklung, Forschungsbericht 4, Hochschulgruppe Arbeits- und Betriebsorganisation HAB e.V., gmft-Verlag, München, 1992, p. 123-159

- Grabowski, H.; Schmidt, M.: Verteilte Konstruktion : Das Arbeiten in Konstruktionsräumen, CAD´92, Neue Konzepte zur Realisierung anwendungsorientierter CAD-Systeme, GI-Fachtagung, Berlin, 14.-15. May 1992, Springer-Verlag, p. 219-232

- Mayer, R. W.: CACID Computer aided concurrent integral design, proceedings of the IFIP TC5/WG 5.7, international working conference on new approaches towards one-of-a-kind production, 1991

- Schmidt, R. F.: Concurrent design - Verkürzung von Entwicklungszeiten durch paralleles Konstruieren, Springer-Verlag, Konstruktion 45 (1993), p. 145-151

- Schmidt, R.F.: Concurrent design - the first step towards simultaneous engineering, proceeding of the 9 th CIM-Europe annual conference, 12.-14. May 1993, Amsterdam (NL)

Research Reports ESPRIT

Area *Computer-Integrated Manufacturing and Engineering (CIME)*

Improving the Performance of Neutral File Data Transfers. Edited by R.J. Goult, P.A. Sherar. IX, 138 pages. 1990 (Project 322 CAD*I, CAD Interfaces, Vol. 6)

Advanced Modelling for CAD/CAM Systems. Edited by H. Grabowski, R. Anderl, M.J. Pratt. VI, 113 pages. 1991 (Project 322 CAD*I, Vol. 7)

IMPPACT Reference Model. Edited by W.F. Gielingh, A.K. Suhm. XII, 261 pages. 1993 (Project 2165 IMPPACT, Integrated Modelling of Products and Processes using Advanced Computer Technologies)

CIMOSA: Open System Architecture for CIM. Edited by ESPRIT Consortium AMICE. XI, 234 pages. 2nd, rev. and ext. edition 1993 (Project 688/5288 AMICE, A European CIM Architecture)

Vibration Control of Flexible Servo Mechanisms. Edited by J.-L. Faillot. VII, 206 pages, 1993 (Project 1561 SACODY, A High Performance Flexible Manufacturing System (FMS) Robot with On-Line Dynamic Compensation)

CCE: An Integration Platform for Distributed Manufacturing Applications. A Survey of Advanced Computing Technologies. Edited by ESPRIT Consortium CCE-CNMA. XII, 207 pages. 1995 (Project 7096 CCE-CNMA, CIME Computing Environment: Integrating CNMA, Vol. 1)

MMS: A Communication Language for Manufacturing. Edited by ESPRIT Consortium CCE-CNMA. XII, 185 pages. 1995 (Project 7096 CCE-CNMA, Vol. 2)

Subseries PDT (Product Data Technology)

CAD Geometry Data Exchange Using STEP. Edited by H.J. Helpenstein. XIV, 432 pages. 1993 (Project 2195 CADEX, CAD Geometry Data Exchange)

Neutral Interfaces in Design, Simulation, and Programming or Robotics. Edited by I. Bey et al. XV, 334 pages, 6 figs. 1994 (Project 2614/5105 NIRO, Neutral Interfaces for Robotics)

NEUTRABAS. A Neutral Product Definition Database for Large Multifunctional Systems. Edited by H. Nowacki. XII, 203 pages. 1995 (Project 2010 NEUTRABAS)

Computer Aided Concurrent Integral Design. Edited by R.F. Schmidt and M. Schmidt. X,78 pages. 1996 (Project 5168 CACID)

Springer-Verlag
and the Environment

We at Springer-Verlag firmly believe that an international science publisher has a special obligation to the environment, and our corporate policies consistently reflect this conviction.

We also expect our business partners – paper mills, printers, packaging manufacturers, etc. – to commit themselves to using environmentally friendly materials and production processes.

The paper in this book is made from low- or no-chlorine pulp and is acid free, in conformance with international standards for paper permanency.

Printing: Saladruck, Berlin
Binding: Buchbinderei Lüderitz & Bauer, Berlin